Nostr

Also from EATMS Productions

Books on power, survival, women's autonomy, and the systems shaping modern America.

Nonfiction

Analysis of Authoritarian Project 2025

Project 2025: The Blueprint
Project 2025: The List
Project 2025, Christian Dumb Dumbs, & The Republican Agenda
Fascism, Project 2025, & The Pinkprint

Autonomy

Bitcoin for Women
Nostr for Women
Digital Self Defense for Women

Billionaires, Capitalism, and Power

Evil and the Mountain Ungreed
Self Help for American Billionaires
Selfish Steve and the Ivory Tower
Tariffs, Taxes, & Face-Eating Leopards
Ban Billionaires: Fascism Fix

The Epstein Class

The Procurement System
The Shield System
The Respectability System
The Narrative System
The Condition

Fascism, Religion, and Cultural Control

Self Help for the Manosphere
Fascism 2025: and the winner is
Fascism & the Perverts & the Greed Virus
Christian Fascism Marriage Book
Tyranny, Table Manners, & Tiramisu

Guides for Women's Autonomy and Protection

How to Survive in Post-America as a Woman
Project 2025 American Drag: Women's Study Bible FTP
4B – Burn, Ban, Boycott, Build (4B Movement)
4B OG – So No Go GYN (4B Movement)
I'm Glad He's Dead
How to Become a Christian in 1 Minute

Modern Rewrites for Women

Stoic Principles Reimagined
Siddhartha Reimagined
The Prince Reimagined for Women
The Art of War Reimagined for Women
The Jungle Reimagined
The Constitution Reimagined for Women

Response to Heritage Foundation's New Project 2025:

Saving America by Saving the Family: A Foundation for the Next 250 Years
The Blueprint II
The List II
The Pinkprint II
The Grayprint Inferno

Fiction

Dystopian Stories of Resistance and Collapse

Propaganda Paige & the Missing Prosperity
Propaganda Paige & the TIDE Manifesto
Propaganda Paige & the Shadow Cartographers
Propaganda Paige & the Prosperity Alliance
Propaganda Paige & the Shattered Truth
Propaganda Paige & the Rising TIDE
Propaganda Paige & the Last Bastion
Propaganda Paige & the Dawn of Prosperity
Project 2025: Dorian — The Last Men
Project 2025: Boy — A Last Men Novel

Survival Guides
For Women in Authoritarian America

Civics
How American Government Actually Works
Parties, Ideologies, & Promises vs Outcomes
Elections, Voting, & Leverage
Power, Money, Media, & the Courts

Infrastructure
Power & Heat
Water & Waste
Food & Supply
Shelter & Movement

Machine Learning
AI, Bitcoin, Nostr for Women
AI, Safety, & Security for Women
AI, Anxiety, & Health for Women
AI, Kids, & Family Safety for Women
AI, Creativity, & Personal Expression for Women
AI, Independent Work, & Parallel Power for Women

Religious Authority as a Governing System
Authority Without Consent
Law as Revelation
The Family as Enforcement Unit
Education, Indoctrination, & Withdrawal
Economics of Submission
Theocratic Convergence

Social Systems
Emotional Labor for Women
Household Power for Women
Workplace Power for Women
Medical Bias for Women
Aging Systems for Women
Recovery Systems for Women

Nostr for Women

The Autonomy Series
Book 2

by
Mads Duchamp

EATMS
PRODUCTIONS

This title is part of an ongoing body of work. All EATMS Productions titles, across all series, authors, and formats, are components of a single connected project.

This book is a work of opinion and creative interpretation. While some names and events may be referenced or alluded to, any claims made are based on publicly available information and are intended as satire, parody, or commentary on societal and political issues. The content should not be interpreted as factual assertions about any individual or entity. The author does not intend to defraud, defame, or mislead, and encourages readers to form their own conclusions. Any resemblance to real persons, living or dead, is purely coincidental unless explicitly noted otherwise.

ISBN: 978-1-966014-46-1

Cover, interior design by: Esme Mees

eatms@pm.me
www.eatms.me

Check out EATMS Underground:
https://tinyurl.com/eatmsNOSTR

Printed in the United States of America.

Freedom of expression begins where permission ends.

— Emma Goldman

Table of Contents

Foreword

Speech shapes women's lives long before anyone explains how it works. It determines what she can say without consequence, how visible she is allowed to be, whether she can warn others, and how easily she can find community when things turn unstable. Yet most women are taught to treat public voice as casual and protected by default. Censorship is framed as moderation. Over time, that confusion becomes a quiet form of control, especially inside systems that reward compliance and punish deviation without explanation.

Nostr enters this landscape not as a trend, but as a contrast. It exposes how much modern speech depends on permission, surveillance, and intermediaries that can decide what is acceptable without accountability. For women, those layers matter. Posts can be buried. Accounts can be restricted. Communities can be erased. Rules can change midstream. In high control environments, whether corporate or governmental, speech systems are often designed to archive everything and never reset the record. Nostr does not remove risk or responsibility, but it changes who holds authority.

This book focuses on understanding that shift. It examines how online speech is managed, who controls visibility, and how those dynamics affect women across platforms and systems. The goal is not ideology. It is to make the structure visible so choices are deliberate rather than assumed.

—Mads Duchamp, Winter 2026

~1
Who Decides What Can Be Said

The Constraint

Most women grow up believing speech is simple. You speak, people listen, and if they disagree, they disagree. The stakes feel social, not structural. The idea of "free speech" is treated like a background fact, something stable and guaranteed. But modern speech does not live in open air. It lives inside systems. It moves through platforms, payment processors, app stores, hosting companies, search engines, and algorithms that decide what gets seen and what gets buried. Speech is no longer just expression. It is access.

For women, that difference matters. Women do not only face disagreement online. They face enforcement. They face misinterpretation, mass reporting, pile-ons, shadow restrictions, and policy shifts that land hardest on anyone who speaks clearly about abuse, power, sex, religion, politics, money, or violence. A man can be blunt and get called brave. A woman can be blunt and get called dangerous. The label changes everything. The system treats her differently, even when the words are the same.

Platforms do not present themselves as governments, but they govern speech in practice. They set the rules, define the punishments, and enforce them without due process. Most of the time you will never know what rule you violated. You will simply notice your reach collapsing. You will notice your posts stop traveling. You will notice the comments dry up. You will notice your account is "limited," "under review," "temporarily restricted," or locked behind a verification demand that feels like a toll booth. The platforms call this safety. It functions like control.

The most important shift is that speech has become conditional. You can speak as long as you behave in ways the system rewards. You can speak as long as you fit into categories that are easy to manage. You can speak as long as you do not trigger a moderation keyword, an automated threat score, or a targeted reporting campaign. In that environment, women learn to self-edit long before they are officially censored. They learn to soften everything. They learn to add disclaimers. They learn to speak in coded language and hope their intended audience understands. That is not freedom. That is adaptation under pressure.

This is not only about politics. It is about platform economics. Social media companies sell stability to advertisers. They sell predictability to investors. They sell compliance to regulators. That means anything that looks chaotic, controversial, or difficult to categorize becomes a liability.

The systems are not designed to protect truth. They are designed to reduce risk. Women's speech often looks "risky" to systems that do not understand context. If she is angry, she is labeled unstable. If she is sexual, she is labeled inappropriate. If she is critical, she is labeled hateful. If she is describing abuse, the system may interpret her as promoting it.

The most dangerous part is that these judgments happen invisibly. A platform rarely has to ban you outright. It can simply change the conditions around your speech so it becomes ineffective. Your posts might technically still exist, but they stop reaching people. Your words might remain on your profile, but they disappear from search. Your account might stay open, but it becomes an isolated room where you speak to yourself. This is what soft censorship looks like. It does not announce itself. It makes you doubt your own perception.

The environment also trains women to build dependence. You are pushed to build your identity, your network, your work, and your credibility inside private systems you do not control. Your name becomes tied to a handle. Your reach becomes tied to an algorithm. Your livelihood becomes tied to engagement metrics. Your ability to be found becomes tied to a recommendation engine. And because these platforms feel like public squares, women often forget the obvious truth: they are corporations. They are not obligated to protect you. They are not obligated to remain consistent. They are not obligated to let you stay.

That dependence creates leverage for anyone who wants to silence you. It can be the platform itself, responding to political pressure or advertiser concerns. It can be a coordinated group, using reporting tools as weapons. It can be an employer, a school, or a government agency that expects compliance. It can be a partner or family member with access to your accounts. Once your voice is routed through a permission system, it can be interrupted. That is the point of the system. It gives someone, somewhere, the ability to say no.

Women also face a second constraint: visibility as exposure. The same systems that suppress speech also amplify it at the worst times. A woman can speak to her community for years and reach almost nobody. Then one post is taken out of context, screenshotted, and pushed into a hostile audience overnight. The platform may do nothing to protect her. It may even reward the outrage because outrage drives engagement. In those moments, speech becomes danger. Not because she did something wrong, but because she spoke in a place designed to monetize conflict.

This creates a trap. If you stay quiet, you disappear. If you speak too loudly, you get targeted. If you speak in the wrong tone, you get punished. If you speak in the right tone, you may still be punished because the system's rules are not consistent. The boundaries are always moving, and the enforcement is always uneven. Women are expected to navigate this with perfect intuition, as if the rules are common sense. But they are not. They are private policies enforced by automation and bureaucracy.

There is also the deeper constraint underneath the platform layer: identity. Most mainstream systems require an identity that is legible to institutions. They want phone numbers, emails, government documents, biometric scans, linked accounts, payment methods, and verified profiles. The more "trust" the system demands, the more it asks you to expose yourself. For women, this is not minor. The demand for proof can turn into a demand for vulnerability. Once your identity is tethered to your speech, consequences become easier to impose. A platform can erase your account and take your history with it. A government can monitor patterns and create dossiers. A hostile person can connect online activity to real-life location and routine.

And once the system has your history, it does not forget. Screenshots are forever. Logs are forever. Search archives are forever. The culture moves on, but the data remains. That is why speech systems shape safety. When speech is attached to identity, it becomes trackable. When it becomes trackable, it becomes governable. And when it becomes governable, women become easier to manage.

The constraint is not that women lack voices. The constraint is that women are expected to speak inside systems designed to control them. Most platforms offer permission-based speech: conditional, monitored, and reversible. You can be present for years and still lose everything overnight. Your audience can vanish. Your work can vanish. Your identity can be weaponized. And the system will call it policy.

That is the reality we begin with. Because before you can protect your voice, you have to see who currently owns the switch.

The Shift

Once you see that modern speech is permission-based, the next question becomes unavoidable. What does it look like to speak without asking? Not in a reckless way, not in a fantasy way, but in a structural way. If the constraint is that platforms quietly govern women's voices, then the shift is to move your voice onto infrastructure that does not rely on a central authority to approve your presence.

That is what Nostr offers. It is not a company, and it is not a single platform. It is a protocol, a way to publish and connect that does not require you to hand over your identity to an institution that can revoke it. Nostr does not promise safety, popularity, or cultural protection. It offers a different architecture. And architecture matters more than intentions, because it determines who holds leverage when the rules change.

Most women do not realize how much modern speech depends on chokepoints. The internet feels open, but the paths that most speech travels through are narrow. App stores can block distribution. Hosting companies can remove access. Payment processors can cut off income. Social platforms can control reach. Search engines can bury links. Even when none of this is called censorship, it functions as control. It sets the boundaries of what is possible and teaches women what happens when they step outside them.

The modern speech environment is built on accounts that are not really yours. Your identity is held by a company. Your reach is filtered through algorithms you cannot inspect. Your audience exists inside a database you do not control. Your history can be removed with one enforcement decision or one policy update. Most of the time, women accept this as the price of participation. But the price keeps rising, and the contract is never stable.

Nostr shifts that balance by separating identity from the platform. On Nostr, your identity is a keypair you control. You are not "approved" by a company. You do not have to ask a corporation to let you exist. Your public key is your identity across the network, which means you can use different apps without starting over. You can move interfaces without losing yourself. That is the difference between an account and infrastructure.

This matters because most modern censorship is not dramatic. It is not always a ban. It is throttling. It is isolation. It is an account that technically exists but no longer reaches anyone. It is visibility reduced until you stop speaking. Nostr does not remove disagreement or conflict, but it reduces institutional leverage. A hostile group can attack your reputation, but they cannot easily erase you everywhere. A client can block you, but it cannot delete your identity. A relay can refuse to carry your posts, but it is only one relay. The point is not that nothing can go wrong. The point is that control is not concentrated in one place.

Women feel the impact of concentrated control first because women's speech is treated as higher-risk by systems that do not understand context. Direct language becomes "aggressive." Descriptions of abuse become "harmful content." Sexual autonomy becomes "explicit material." Criticism becomes "hate." The system does not need to be intentionally misogynistic to create misogynistic outcomes. It only needs to be designed for risk reduction, scale, and automation.

The shift Nostr offers is not only about publishing. It is about continuity. When women lose accounts on mainstream platforms, they do not just lose posts. They lose their networks, their work, their social ties, and the ability to coordinate. They lose access to communities they helped build. The system makes those losses feel personal, as if women failed to play the game correctly. But it is structural. A permission system always reserves the right to revoke.

Nostr makes connection more durable because it is not tied to a single corporate database. Your identity can move between clients. Your social graph can move with you. You can choose different relays. You can rebuild without restarting from zero. That reduces the fragility that platforms depend on to keep people compliant. When women know they have an exit, they behave differently. They stop shaping every sentence around fear of being removed.

In stable times, this might seem unnecessary. In unstable times, it becomes essential. The tightening of control usually happens through existing systems long before new rules are announced. Pressure is applied quietly. Enforcement becomes inconsistent. People are punished as examples. Women who rely on a single platform for visibility are easier to manage because they have nowhere to go. Women who treat speech as infrastructure are less likely to be caught off guard. They are not immune to pressure, but they are less disoriented by it. Disorientation is one of the most effective tools of control.

Nostr does not eliminate risk. No communication system can. What it offers is a different distribution of authority. It gives women more control over identity, more resilience in connection, and more freedom to move. It reduces the number of gates that can be closed at once. It does not guarantee an audience, but it makes it harder for an audience to be taken away by policy.

This shift also changes how women relate to speech itself. Corporate platforms train women to perform. They reward the right tone, the right kind of harmlessness, and the right kind of self-censorship. Over time, women learn to speak for the algorithm instead of speaking for the people they want to reach. They learn to soften the truth. Nostr will not magically remove social consequences, but it makes a different posture possible. You can build a voice based on consistency rather than constant optimization.

The shift is not about abandoning everything familiar. It is about reducing dependency. Most women will still use mainstream platforms for visibility, convenience, and reach. This book does not ask you to disappear from those spaces. It asks you to stop treating them as permanent. The practical strategy is to build a parallel channel where your identity and connections are not owned by someone else. When you do that, a sudden policy change becomes an inconvenience, not a collapse.

As environments harden, this becomes more than a preference. It becomes a survival advantage. When speech is tied to identity, consequences become easier to impose. When speech is routed through corporations, pressure becomes easy to apply. When communication is centralized, coordination becomes easy to disrupt. The shift does not promise safety. No system can. What it offers is clarity. Clarity about where power lives. Clarity about which tradeoffs are real. Clarity about what is being exchanged when convenience is chosen over control.

This chapter does not ask women to be louder. It asks them to be harder to silence. It explains how modern platforms govern speech through invisible enforcement, and why infrastructure like Nostr changes who holds authority over what can be said.

Speech stops being a performance inside someone else's system and becomes what it has always been. A capacity. One that can be protected and carried forward. The shift is not ideological. It is practical. When women understand that speech is never neutral online, they stop expecting platforms to be fair. They stop expecting corporations to defend them. They begin to decide deliberately how much control they are willing to give, and where. That decision is the beginning of autonomy.

~2
The Cost of Being Visible Online

The Constraint

Visibility is sold to women as empowerment. Be seen. Be heard. Build a platform. Share your story. Post your work. Make a brand. The message is constant, and it sounds like freedom. If you want opportunities, you need an online presence. If you want community, you need to participate. If you want safety, you need people to know who you are. But visibility does not operate as a neutral good. It operates as exposure. And for women, exposure has a higher cost.

Men can often be visible online without their bodies becoming the subject. Without their emotions being treated as evidence. Without their relationships being used against them. Women do not get that separation. A woman's speech is rarely heard as "just speech." It becomes a personality test. A morality test. A provocation. A signal. The same post that makes a man look confident can make a woman look unstable. The same disagreement that makes a man look principled can make a woman look "difficult." The penalty is not always public, but it accumulates. Over time, women learn that visibility is not just a spotlight. It is a vulnerability.

The first cost is misinterpretation. Most online systems do not understand context. They understand patterns. A woman speaks about abuse, and the system sees violence. A woman speaks about sexuality, and the system sees explicit content. A woman speaks about anger, and the system sees harassment. A woman speaks about politics, and the system sees risk. Platforms prefer language that is mild, safe, and easily categorized. Women's real experiences are rarely mild. When women speak honestly about their lives, they often trip moderation systems that were designed for convenience, not truth.

The second cost is punishment through visibility itself. The more visible you are, the more likely you are to be targeted. Not because you did something wrong, but because attention attracts predators, opportunists, and bored people looking for a fight. Platforms reward conflict because conflict increases engagement. That means a woman can be pulled into hostile attention simply for being interesting, competent, or direct. She does not need to go looking for trouble. Visibility brings trouble to her.

This is where women learn the difference between "audience" and "crowd." An audience is chosen. A crowd is accidental. Most women want the first and get the second. A woman posts something meant for her small community, and it spreads beyond them. It lands in the feeds of strangers who have no shared context and no investment in her humanity. The platform treats this as success. In real life, it can feel like being surrounded.

The third cost is social punishment. Women are watched more closely. Their tone is graded. Their confidence is treated like arrogance. Their boundaries are treated like hostility. They are expected to be warm, grateful, flexible, and pleasant even when discussing serious topics. If a woman refuses that performance, she is punished socially, professionally, and algorithmically. A man can be blunt and be called a leader. A woman can be blunt and be called a threat.

Visibility also creates documentation. Most women underestimate how permanent the internet is. Posts can be screenshotted, archived, copied, and reshared without your consent. Even if you delete something, the record may already be elsewhere. The culture moves quickly, but the data stays. That permanence changes the meaning of speech. A sentence written in exhaustion can become evidence later. A joke can become a character judgment. A private story can become a weapon. Women are expected to predict the future interpretation of everything they say, and they are punished when they cannot.

The fourth cost is sexualization. Women's visibility is treated as availability. That is not a metaphor. Women who become visible online often receive messages that range from crude to threatening to obsessive. They are watched. They are commented on. They are approached as if their presence is consent. This is not "just the internet." It is a social pattern reinforced by systems that do not meaningfully punish predatory behavior. Platforms allow women to be targeted because outrage and harassment still drive engagement. The woman pays the price. The platform collects the data.

The fifth cost is identity linkage. The more visible you are, the more the world can connect your speech to your real life. Your name, your location, your job, your family, your routines. The internet makes it easy to pull small pieces of information together until a stranger knows too much. This is not paranoia. It is the logic of open-source intelligence. Women are not imagining it when they feel watched. Many women are watched. And the more visible they become, the easier it is for someone to move from online harassment to offline interference.

This is why women face higher penalties for speaking openly. Their speech is not evaluated only as speech. It is evaluated as permission. As legitimacy. As a signal that can be punished. A woman who speaks about money becomes greedy. A woman who speaks about power becomes bitter. A woman who speaks about safety becomes hysterical. The system does not require an official censor. It only requires a culture and a platform structure that encourages people to police women for existing outside the acceptable range.

Platforms add another layer of punishment: selective amplification. Women often experience the worst of both worlds. Their ordinary work is ignored. Their hardest truths are buried. But the moment something can be used against them, it spreads. A woman can post steady, valuable material for months and receive silence. Then one clipped sentence becomes viral in the wrong direction. She is not rewarded for her value. She is punished for her visibility. And because platforms thrive on volatility, they are rarely motivated to protect her.

This shapes women's behavior over time. They become cautious. They second-guess. They soften. They retreat. They choose silence over risk. Not because they lack courage, but because the cost is real. The penalty is not just emotional. It can be financial. Professional. Relational. Women can lose jobs over online exposure. They can lose housing. They can lose custody battles. They can lose access to accounts and services. A single moment of visibility can create consequences that last for years.

And the most effective part of this system is that it still looks voluntary. Women are told they chose to post. They chose to be public. They chose to "engage." The platform hides behind that framing. But the reality is that modern life pushes women toward online participation, then punishes them for being there. That is the contradiction. Women are expected to be visible in order to be relevant, then shamed and controlled for the vulnerabilities that visibility creates.

This is the constraint: visibility is not evenly priced. Women pay more for the same exposure. They pay in harassment, in misinterpretation, in surveillance, and in reputational risk. They pay in the constant need to manage tone, manage safety, and manage the future consequences of ordinary speech. The system calls this participation. For women, it often functions as a test.

The Shift

Once you understand that visibility is not neutral, the goal changes. The goal is not to be invisible. It is to become visible on your own terms. Women do not need to disappear from public life. They need to stop paying unnecessary costs for visibility inside systems that are designed to extract value from their exposure.

Most mainstream platforms treat women's visibility as content. It is monetized, sorted, and traded. Your attention becomes inventory. Your reactions become data. Your body becomes engagement. Your identity becomes a profile that can be sold, scored, or mapped. This is why being visible online feels unstable. You are not seen as a person first. You are seen as a signal. And signals are managed, not protected.

The shift is to treat visibility as infrastructure instead of performance. Infrastructure means you decide what is attached to your identity, what is shared, and what is kept separate. It means you build a public presence that does not automatically leak your private life. It means you stop using the same identity for everything. Most women were taught that consistency is safety. One name, one profile, one account, one life. In a high-surveillance world, that approach is fragile. It makes you easy to track, easy to pressure, and easy to punish.

Nostr supports a different posture because your identity is not granted by a platform. It is held by you. That does not mean anonymity is automatic, but it means control is possible. You can create identities intentionally rather than being forced into one permanent profile that everything attaches to. You can decide what is public-facing and what is private. You can separate professional work from personal life. You can keep a stable public voice without giving strangers a direct path into your offline world.

This is not about hiding. It is about boundaries. Boundaries are what platforms quietly discourage because boundaries reduce extraction. A woman with strong boundaries is harder to monetize. She is harder to manipulate. She is harder to pressure into performing. Most corporate systems reward women who are open, constant, and emotionally accessible. They punish women who are private, steady, and difficult to provoke. The shift is to stop negotiating your boundaries with algorithms.

The next part of the shift is understanding that harassment thrives on centralized attention. Most targeted attacks require a platform to amplify them. A dogpile works when there is one stage and a shared crowd. Nostr does not remove hostility, but it changes how attention aggregates. It becomes less automatic. It becomes less dependent on one algorithm pushing conflict into millions of feeds. That reduces the likelihood of sudden mass exposure from one clipped post. It also gives you more control over who you see and who sees you through your choices of client, relay, and network.

Women often assume that safety online comes from better rules. Stricter moderation. More enforcement. More reporting tools. In practice, those tools are often used against women, not for them. Reporting systems are easily weaponized. Enforcement is inconsistent. The same platform that fails to stop stalking will quickly punish a woman for being "rude." This is why women cannot outsource safety to policy. The shift is to build systems where safety is supported by structure, not by promises.

Visibility also becomes safer when it is diversified. A woman who relies on one platform has one point of failure. One ban, one policy shift, one coordinated reporting event, and she is erased. When women build parallel channels, they reduce the damage. Nostr becomes a place where you can be found even if something else collapses. It becomes a contact point that is less dependent on the goodwill of a corporation. It becomes a continuity layer.

This shift is practical. It looks like choosing what you publish with your real name and what you publish under a separate identity. It looks like speaking publicly without tying your speech to your home address, your workplace, your children's photos, or your daily routines. It looks like reducing how much your audience can infer about your offline life. Not because you are ashamed, but because you are entitled to safety. Women have been trained to treat privacy like guilt. In reality, privacy is a boundary, and boundaries are a form of power.

It also looks like changing your relationship to engagement. Most platforms reward women for being reactive. For responding quickly. For staying emotionally available. For proving they are "real." But responsiveness is a vulnerability. It gives strangers a lever to pull. It gives mobs a way to drag you into their tempo. Nostr makes it easier to stop chasing the performance loop, because the incentive structure is different. You can be consistent without feeding the machine. You can publish and leave. You can build reputation through steadiness rather than through constant output.

Women also gain safety when they understand how identity linkage works. The internet is not dangerous only because of what is posted. It is dangerous because of what can be connected. A photo can reveal a location. A small detail can reveal a routine. A name can reveal a family network. The shift is to treat this as a design problem, not a moral one. It is not about being paranoid. It is about reducing predictable vulnerabilities. Every woman should assume that anything public can be archived, copied, and used later. That is not fear. That is realism.

Nostr does not solve the entire visibility problem, but it changes the default assumptions. It makes it easier to maintain separate identities. It makes it harder for a single platform to erase your presence. It creates more space for women to build public voice without handing over total control of the channel. And over time, it trains women to treat their online presence as something they actively manage, not something that happens to them.

In high compliance structures, this shift becomes essential. Control tightens through existing systems before new rules are announced. Women who understand visibility as exposure are less likely to be caught off guard. They are not immune to pressure, but they are less disoriented by it. Disorientation is one of the most effective tools of control. The shift does not promise safety. No speech system can. What it offers is control over the variables that can actually be controlled: identity, access, continuity, and boundaries.

This chapter does not ask women to stop being seen. It asks them to stop being exposed by default. It explains why visibility carries higher penalties for women, and why shifting to infrastructure like Nostr helps women build public voice with less institutional leverage and less fragility.

Visibility stops being a performance and becomes what it has always been. A choice. One that can be shaped, limited, and defended. The shift is not ideological. It is practical. When women understand that visibility is never neutral online, they stop expecting platforms to protect them. They stop trading privacy for validation without noticing. They begin to decide deliberately what gets attached to their name, what gets attached to their work, and what stays theirs alone. That decision is the beginning of safety.

~3
Centralized Platforms and Silent Control

The Constraint

Centralized platforms do not need to ban women to control them. They only need to shape the environment in which women speak. Most control online is not loud. It is silent, procedural, and easy to deny. A platform does not have to say, "You are not allowed to speak." It can simply decide what speech travels, what speech stalls, and what speech disappears into a dead zone. That is enough to change behavior.

This is how modern governance works in digital spaces. Instead of laws, there are policies. Instead of courts, there are automated systems and moderation teams. Instead of appeals, there are vague forms and canned responses. The effect is similar: women learn what speech costs, and they adjust. They stop talking about certain topics. They change their tone. They avoid certain words. The platform never has to argue with them. It simply trains them.

The first mechanism of silent control is moderation. Moderation is framed as safety, and sometimes it is. But the way it operates at scale creates predictable harm. Most moderation is not done by humans who understand context. It is done by automation that sees only patterns, keywords, and reports. Even when humans are involved, they are often overworked, undertrained, and pressured to make fast decisions.

This produces a system that favors the simplest outcome: remove, restrict, or suppress anything that looks messy.

Women's lives are messy. Reality is messy. Language that describes actual harm is often flagged as harmful. Women speaking about violence can be treated as violence. Women speaking about sexuality can be treated as pornography. Women speaking about coercion can be treated as "adult content." Women speaking about politics can be treated as extremism. Platforms say they are enforcing rules evenly, but they are enforcing them through blunt instruments. The result is uneven control with a clean corporate face.

The second mechanism is reporting. Reporting is supposed to stop abuse. In practice, it can be weaponized. A coordinated group does not have to refute a woman's ideas. They can mass-report her content until the system triggers restrictions automatically. They can flag her account as dangerous, hateful, or unstable. And because platforms treat reports as signals of risk, the system often reacts first and asks questions later. A woman can lose reach, lose access, or lose her account without ever being told why in a meaningful way.

This is why the experience of being "censored" often feels surreal. There is no argument. There is no clear accusation. There is only the sensation of doors closing. Features stop working. A post doesn't show up. A link is blocked. A video is removed. The account is "limited." Visibility collapses. These are small control actions, individually deniable, but together they shape who can participate in public life.

The third mechanism is algorithmic visibility. Most platforms do not show content in chronological order. They show what the system predicts will keep people scrolling. That means speech is rewarded based on engagement, not on truth, usefulness, or integrity. The algorithm becomes the real editor. It decides which voices feel "relevant" and which voices feel invisible. Women can write steady, valuable material and watch it disappear. Meanwhile, sensational content rises because it creates stronger emotional reactions.

This shapes behavior even when women try to resist it. When your survival depends on being seen, you start learning what gets rewarded. You learn which tones travel. You learn which topics trigger conflict. You learn which formats generate clicks. And slowly, the platform trains you to speak in ways that are compatible with its incentives. Women think they are just "learning social media," but what they are learning is compliance with a profit engine.

This is why the most dangerous control is not restriction. It is incentive. Incentives reshape people from the inside. When a platform rewards outrage, it manufactures outrage. When it rewards vulnerability, it pressures women to perform vulnerability. When it rewards simplistic narratives, it punishes complexity. When it rewards aesthetics, it punishes seriousness. Women who want to be heard are gradually pushed toward the speech that the machine can monetize.

The fourth mechanism is monetization pressure. On many platforms, income is tied to visibility. Creators are rewarded for constant output, constant engagement, constant performance. Women learn that silence costs money. Stepping away costs money. Refusing to play the game costs money. This creates a subtle form of coercion. A woman who needs income cannot always afford to be careful. She becomes dependent on the platform's rules and moods. That dependency is leverage.

Even women who are not monetizing directly are still shaped by the same forces. Visibility is social currency. It affects employment, opportunities, and reputation. When platforms decide what is seen, they decide what is credible. That makes them powerful cultural institutions, even if they pretend to be neutral tools. A platform can quietly decide which topics are "dangerous," which communities are "acceptable," and which women are "too much." It does this without announcing a political stance. It simply tunes the machine.

The fifth mechanism is friction. Centralized platforms can make certain actions easier and others harder. They can make sharing links inconvenient. They can hide external content. They can push users toward features that increase platform lock-in. They can nudge speech toward short, emotional bursts instead of long, thoughtful explanations. This is not accidental. It is design. The platform shapes the kind of speech that thrives, and the kind that dies.

For women, these mechanisms create a specific pattern of control: speech is allowed as long as it stays within a narrow range of acceptable femininity. A woman can be visible if she is entertaining, attractive, soft, and non-threatening. She can be political if she is harmless. She can be serious if she is polite. She can be angry if it is framed as humor. When she steps outside those boundaries, the system becomes less generous. It does not always punish her openly. It simply removes the rewards.

The constraint, then, is not only censorship. It is behavioral shaping. Women are not simply being silenced. They are being trained. They are being nudged toward safer speech, smaller speech, softer speech, less effective speech. They are being encouraged to trade clarity for acceptance.

And because the control is silent, it is easier to internalize. Women start believing they are the problem. Women start believing they just need to be "better online." They do not realize the platform is working as designed.

The Shift

Once you understand how silent control works, the goal changes. The goal is not to argue with centralized platforms or hope they become fair. The goal is to reduce how much power they have over your voice. Centralized platforms are not broken versions of public squares. They are working systems built to generate profit, reduce liability, and maintain leverage. When women accept that, they stop treating platform outcomes as personal failures and start treating them as structural constraints.

The shift begins with a simple change in mindset: moderation is not the only form of control. Control also happens through incentives. The platform does not need to silence you. It only needs to reward speech that is safe for the business model and punish speech that creates risk. That is why women can feel "allowed" to speak while still being trained into smaller, softer, less effective language. A system can permit expression while still shaping behavior.

Nostr changes this because it is not built around a central incentive engine. It does not have a single algorithm deciding what deserves visibility. It does not have advertisers demanding a sanitized environment. It does not have a corporate reputation to protect. That does not mean it is clean or comfortable. It means the control structure is different. The rules are not unified under one authority, which makes silent governance harder to impose at scale.

The most practical shift Nostr offers is distribution. On centralized platforms, there is one owner and one set of invisible levers. On Nostr, your identity is not owned by a company. Your account is a keypair you control. Your presence is not dependent on approval from a platform's trust system or verification pipeline. That changes the relationship between speech and permission. You are not renting your voice from a corporate landlord. You are carrying it.

This matters because so much platform control relies on the threat of removal. When women believe their entire digital life lives inside one company's boundaries, they become easier to manage. They soften. They perform. They self-censor. They avoid conflict. They trade clarity for continuity. A woman does not need to be banned to be controlled. She only needs to believe she cannot afford to lose access. Nostr reduces that leverage by making identity portable. When women can move, they become harder to trap.

The shift also changes how women relate to visibility. Centralized platforms convert attention into a currency they control. They decide what becomes visible and what becomes invisible. They make visibility feel random, then sell the illusion that it can be "earned" through better behavior, better branding, or better compliance.

Nostr does not eliminate attention dynamics, but it removes one major source of manipulation: a single recommendation engine whose purpose is to maximize time-on-platform. Without that engine, speech is less optimized for addiction and more capable of being informational.

Another shift is in how enforcement works. On centralized platforms, enforcement is opaque and centralized. You are punished by a system you cannot inspect or negotiate with. On Nostr, enforcement is fragmented and local. One client might block content. One relay might refuse to host certain posts. One community might reject you socially. These are still constraints, but they are not unified under a single authority. That means women have options. Options matter more than perfect safety, because options allow recovery.

This is the practical difference between centralized governance and decentralized resilience. Resilience is not a claim that nothing bad happens. It is a design that prevents one event from becoming total collapse. A woman on a centralized platform can be erased overnight. A woman on Nostr can be disrupted, criticized, filtered, or blocked in specific places, but she is harder to erase everywhere at once. That changes the consequences of speaking. It reduces the fear that one mistake, one controversy, or one targeted campaign ends everything.

The shift also requires women to become more deliberate about identity and boundaries. Centralized platforms often encourage total exposure because exposure is profitable. They train women to merge personal and professional life into one legible, trackable profile. Nostr supports a different approach because you can hold multiple keys and separate contexts. You can decide what voice is attached to what identity. You can build a public persona that does not automatically leak your private life.

Nostr also helps women resist incentive capture. When you are no longer speaking to an algorithm, you stop shaping your voice around what the machine rewards. You can write longer thoughts without being punished for not being entertaining. You can share serious information without turning it into rage bait. You can build trust through consistency instead of popularity spikes. That does not guarantee influence, but it restores dignity. It allows speech to function as communication again instead of constant performance.

The shift is not to abandon every centralized platform overnight. The shift is to stop treating centralized platforms as permanent foundations. Use them if you want, but do not build your entire voice inside them. Build parallel infrastructure. Create a contact point that is harder to delete. Create a network that can survive policy shifts. Create a public identity you can carry across clients and communities. This is the same logic as financial redundancy. You do not rely on one bank. You do not rely on one channel.

In authoritarian contexts, this shift becomes essential. Pressure increases through existing systems before new rules are announced. Incentives become sharper. Enforcement becomes political. Speech becomes a liability. Women who understand how platforms shape behavior are less likely to be caught off guard. They are not immune to pressure, but they are less disoriented by it. Disorientation is one of the most effective tools of control. The shift does not promise freedom. No network can. What it offers is leverage. Leverage through portability. Leverage through optionality. Leverage through reduced dependency.

This chapter does not ask women to fight centralized platforms. It asks them to see how silent control actually operates through moderation, algorithms, and incentives, and why moving speech onto infrastructure like Nostr reduces the platform's ability to shape behavior without consent.

Speech stops being something you perform under surveillance and becomes what it has always been. A capacity. One that can be protected, distributed, and sustained. The shift is not ideological. It is practical. When women understand that platforms are not neutral and incentives are not harmless, they stop begging for fairness. They stop expecting consistency from systems designed to change. They begin to design their communication in ways that preserve autonomy even when the environment tightens. That decision is the beginning of durability.

~4
What Nostr Is (Without the Hype)

The Constraint

Most women first hear about Nostr the same way they hear about everything else online: through noise. Someone calls it the future. Someone calls it a cult. Someone says it will "fix social media." Someone says it is only for Bitcoin people. The tone is usually extreme, and the explanations are usually useless. That is the first constraint. Nostr is often introduced with hype, jargon, and insider language, which makes it feel inaccessible on purpose.

Women are already exhausted by tech that demands attention and gives little clarity. Most platforms require you to learn their unspoken rules, absorb their cultural moods, and adapt constantly. The last thing women need is another system that feels like a test. If Nostr is explained as a movement, a vibe, or an ideology, many women will ignore it. Not because they are incapable, but because they are tired of being recruited into someone else's story.

The second constraint is that people confuse apps with infrastructure. Most users think the platform is the product. Twitter is Twitter. Instagram is Instagram. You download the app, create the account, and the company owns the environment. That mental model is so common that it becomes invisible. It is also why centralized platforms feel permanent until they collapse.

When women hear "Nostr," they assume it is another app. They search for it in an app store and expect one official version. But Nostr is not that. It is a protocol. And protocols feel unfamiliar because they are not marketed.

This is where confusion starts. A protocol is not a brand. It is a set of rules that different tools can use to communicate with each other. Email is a protocol. The web is a protocol. You can use different browsers to access the same websites. You can use different email clients to send messages through the same system. No single company owns email, even though companies can build products on top of it. Nostr works more like that. It is a way for messages to move across a network without one platform controlling the whole thing.

But because modern platforms trained women to expect centralized ownership, this can be hard to picture. Women are used to a single account, a single company, a single feed, a single set of rules. Nostr breaks that model, and anything that breaks the familiar model creates discomfort. Discomfort is the constraint. Women do not need more discomfort. They need a simple explanation that respects their time.

The third constraint is technical intimidation. Many explanations of Nostr lead with words like relays, clients, keys, signatures, and encryption. Those are real concepts, but they are not the right place to start. Starting with jargon makes women feel behind before they begin. Most women have learned that tech language is used as a gate. It signals who belongs and who does not. Even when the community does not intend harm, the effect is the same. Women either pretend they understand or they leave.

The fourth constraint is trust. Centralized platforms are awful, but they are familiar. Women know what they are getting. They know how to make an account, how to post, how to block people, how to leave. Nostr requires a different trust posture, because the responsibility shifts toward the user. If you lose your keys, you lose your account. If you choose a bad client, you will have a bad experience. If you join the wrong relay culture, you will feel overwhelmed. That can sound risky, especially to women who already carry risk everywhere.

The truth is that the responsibility shift is the point. Centralized platforms feel easy because the platform carries control and surveillance for you. It holds your identity. It controls your access. It resets your password. It can recover your account. That convenience comes with a trade: the platform can also remove you, throttle you, or monetize you without your consent. Nostr does not offer the same kind of convenience because it does not offer the same kind of control. And women have to understand that trade clearly to choose it.

The fifth constraint is that people treat decentralization as moral superiority. They speak as if decentralization automatically makes a system good. It does not. Decentralization is not a virtue. It is a structure. It changes who holds power and how failure works. It can reduce censorship leverage, but it can also feel messier. It can be harder to moderate. It can be harder to police harassment. Women do not need propaganda. They need honesty. They need to know what decentralization actually changes and what it does not.

So here is the simple picture. Nostr is a way to publish messages to a network using an identity you control. You create a keypair. The public key is how people find you. The private key is how you prove you are you. You use an app called a client to write posts and read posts. Your client connects to servers called relays, which pass messages around. Relays do not own your identity. They are more like bulletin boards. Some are better than others. Some are open. Some are curated. Your client can connect to multiple relays at once, which means you are not dependent on one place.

That may still sound technical, but the idea is simple. Your identity is portable. Your posts are distributed. No single company owns the network. If one relay disappears, you can use another. If one client becomes unusable, you can switch. Your identity stays the same because it is tied to your keys, not to an app's database.

This is the part women need most: you can think of Nostr like email for social media. With email, you do not lose your identity because one company changes its rules. You can choose a provider. You can move. You can use different apps to read and send messages. Nostr is similar. It is a base layer that multiple tools can use. That means it is harder for one entity to control the whole system.

And because it is a base layer, Nostr is not limited to one type of communication. It can support short posts, longer writing, messaging, community spaces, and even identity and payment tools layered on top. The system is flexible because the protocol is simple. That is also why it attracts both serious builders and loud personalities. Women do not have to care about the culture war around it to benefit from the structure. They only need to understand what it gives them: portability, optionality, and fewer centralized points of control.

The constraint is that most women have been trained to see technology as a product, not as infrastructure. They have been trained to hand over identity for convenience and call it normal. They have been trained to accept algorithmic feeds as reality. Nostr breaks those defaults. And until women understand what it is in simple terms, it will keep getting dismissed as hype, niche, or too technical.

The Shift

Once Nostr is understood as infrastructure instead of a trend, the decision becomes easier. You stop asking whether you "like the community" or whether the vibe matches your personality. You start asking a more useful question: what kind of communication system do you want your life built on. Most women are used to treating platforms like places. Nostr is not a place. It is a structure that lets you build places without surrendering your identity to a single owner.

The shift begins by letting go of the app mindset. On centralized platforms, the app is everything. The company owns the identity system, the social graph, the algorithm, the rules, and the data. You exist inside their walls. On Nostr, the app is just an interface. It is a client. You can choose one and switch later. Your identity does not belong to the client. Your relationships do not belong to the client. Your posts are not trapped inside one company's database. That portability is the core difference, and it is why Nostr is worth learning even if you never become a "power user."

The next shift is understanding keys without fear. Your public key is your identity. Your private key is your control. Centralized platforms replace this with passwords and recovery systems because they want to hold your identity for you. That creates convenience, but it also creates dependency. Nostr does the opposite. It gives you direct ownership, which means you gain autonomy and also gain responsibility. For women, this is not a burden. It is a trade that can be managed with a few disciplined habits. You do not need to understand cryptography. You need to understand custody. Your private key is your account. Protect it the way you would protect access to money.

Relays are the next piece. Relays are often described as if they are mysterious. They are not. They are servers that receive and share posts. You can use many at once. You can add new ones. You can remove ones that are unreliable or hostile. This creates redundancy. Centralized platforms do not offer redundancy because redundancy weakens their control. If you depend on one company, that company can shape your behavior. If you can route around it, you become harder to manage.

This is the shift: Nostr replaces dependence with options. Options are what women lose first in tightening environments. When platforms become stricter, when policies become political, when enforcement becomes uneven, women discover they have no alternatives. They are trapped inside one system because they built everything there. Nostr offers a different approach. You can build a parallel channel that survives changes elsewhere. You can establish a contact point that is harder to erase. You can create continuity that does not depend on corporate permission.

The shift also changes what "visibility" means. On centralized platforms, visibility is produced by algorithms and purchased by advertisers. It is unstable by design. You can be visible one day and invisible the next because the platform's incentives changed. Nostr is not algorithm-driven in the same way. It is closer to a network of direct connections. That means growth can feel slower, but it is often more real. You are not performing for a machine. You are connecting to people. The tradeoff is that attention becomes less automatic and more intentional, which is healthier for women who are tired of being trained into performance.

This matters because most women do not want fame. They want durable reach. They want to be able to publish information, stay connected, and find community without being punished by policy shifts. Nostr is not a guarantee of safety, but it reduces one major risk: the ability of a single platform to cut you off from your own identity. That alone changes the emotional temperature of speech. When you can leave, you do not have to beg.

Nostr also changes the terms of moderation. Centralized platforms promise safety through enforcement, then fail women at scale. They punish women for tone while letting harassment thrive. They enforce rules unevenly while claiming neutrality. Nostr does not offer one unified safety system, but it does offer a different kind of control: user-level control. You can curate your experience. You can choose who you follow. You can choose clients with stronger filtering tools. You can choose communities that align with your goals. This does not eliminate conflict, but it shifts control closer to you.

The shift is not to pretend decentralization is perfect. It is to recognize that centralized systems have structural incentives that will never prioritize women. They will prioritize advertisers, regulators, and shareholders. If women want a communication system that is harder to manipulate, they need one that is harder to capture. Nostr is not immune to capture, but it is less centralized, which means capture is slower and less total. That is what resilience looks like.

On a practical level, the shift looks like building competence in small steps. You choose a client that feels usable. You save your keys securely. You connect to a few relays. You follow a handful of people whose posts are worth reading. You learn what good signal looks like. You build your feed slowly. You do not try to recreate the entire mainstream platform experience in a week. Nostr rewards patience because it is infrastructure, not entertainment.

It also helps to treat Nostr like a publication tool, not a lifestyle. Women do not need to adopt anyone's ideology to benefit from protocol-based communication. They can use Nostr the way they use email: as a reliable channel. A place to post writing, announce work, share links, and maintain contact with people who matter. The more women treat it as infrastructure, the less they get dragged into personality wars around it. The structure is the value, not the drama.

In closing systems, the simplicity becomes the advantage. Restrictions increase through existing systems before new rules are announced. Platforms quietly adjust visibility. Banks and payment processors quietly restrict access. Institutions quietly pressure compliance. Women who rely entirely on centralized networks are easier to isolate. Women who build parallel channels are harder to silence completely. They are not immune to pressure, but they are less disoriented by it. Disorientation is one of the most effective tools of control. The shift does not promise safety. No network can. What it offers is continuity. Continuity of voice. Continuity of identity. Continuity of connection.

This chapter does not ask women to become technical. It asks them to see Nostr clearly: as a simple protocol for decentralized communication, where identity is portable and control is not concentrated in one company. The shift is practical, moving from app dependence to infrastructure that can be carried forward.

Communication stops being a service you rent and becomes what it has always been. A capacity. One that can be structured, defended, and kept intact across changing conditions. The shift is not ideological. It is practical. When women understand that a platform is not the internet, they stop confusing convenience with safety. They stop treating corporate permission as normal life. They begin to build communication systems that do not collapse the moment a gate closes. That decision is the beginning of resilience.

~5
Identity Without a Platform Owner

The Constraint

Most women are taught to measure identity online through status. Followers. Likes. Verification badges. Visibility. A clean profile. A consistent brand. The assumption is that identity is something you build inside a platform and then the platform reflects it back to you as "credibility." Women are told this is how the internet works. You establish yourself. You gain traction. You become "known." But the system underneath that story is not about identity. It is about ownership.

On centralized platforms, identity is rented. You may feel like your account is "yours," but it is not. The platform controls the database. It controls authentication. It controls recovery. It controls your visibility. It controls whether your profile appears in search. It controls whether your links work. It controls whether your content is considered acceptable.

A centralized platform can reshape your identity without ever touching your real life. It can make you invisible. It can label you risky. It can limit your reach. It can remove your archive. It can erase your social graph. It can do all of this in a way that looks administrative, not political.

Women often do not recognize this as identity control because it is disguised as customer service. You are "locked out." Your account is "under review." You violated a policy. You need to verify again. Your content is "not recommended." Your reach is "reduced for safety." The language is soft. The consequence is hard. Identity becomes conditional. You are not you because you are you. You are you because the platform currently allows you to be.

This is one of the most important constraints of modern life. Women are being trained to tie their identity to systems that can revoke it. And because women often rely on networks for safety, work, and emotional support, the cost of losing an account is high. It is not just inconvenient. It can cut off income. It can cut off community. It can cut off information. It can cut off social proof that took years to build. When women know that, they become easier to manage.

This is why follower counts become a trap. Followers feel like ownership, but they are not. They are numbers inside a corporate database. You cannot export them meaningfully. You cannot contact them directly without the platform's permission. You cannot ensure your posts reach them. A platform can quietly reduce distribution while still showing you the follower number, which keeps you attached to the illusion that you still have something valuable. It is a control technique. The platform makes you feel invested so you will tolerate instability.

Women also face a deeper risk: identity linkage. Most centralized platforms push toward real-name systems, verified identities, phone number connections, and cross-platform integration. This is framed as trust and safety. Sometimes it is about fraud prevention. But it is also about legibility. Institutions want users to be legible because legible people are easier to monitor, predict, and pressure. The more your identity is anchored to institutional proof, the harder it is to speak freely. Consequences become easier to apply.

For women, identity linkage is not abstract. It increases exposure to stalking, harassment, and offline retaliation. It makes it easier for employers, family members, or hostile strangers to connect online speech to offline life. It makes reputation attacks more effective because the target is easier to locate. It also makes the emotional cost of speaking higher, because speech becomes tied to social punishment in every environment at once. Women are expected to carry that risk quietly while still participating as if it is normal.

Centralized identity also creates dependency through recovery. Platforms encourage women to believe that the system will "help" if something goes wrong. Password resets. Account recovery forms. Verification processes. The truth is that recovery is a control function. It determines who is allowed to reclaim an identity, and under what conditions. Recovery systems can be used to protect users, but they can also be used to pressure compliance. A platform can demand more data, more proof, more submission. A woman who depends on the account will provide it.

There is also the issue of reputation as a commodity. Centralized platforms package identity into a marketable object. Your account becomes a product. Your persona becomes content. Your relationships become engagement. Women are encouraged to perform identity constantly, to stay relevant, to stay visible, to stay aligned with the platform's incentives. Over time, women stop distinguishing between who they are and what the platform rewards. They begin to think of identity as something that can be optimized. That is a subtle kind of distortion.

This leads to another constraint: identity becomes fragile when it depends on constant approval. Women feel pressure to stay within a safe range of acceptable speech and acceptable presentation. They learn to keep the tone friendly. They learn to avoid sharp edges. They learn to pre-apologize. They learn to act grateful for attention, even when the attention is exploitative. This is not because women are weak. It is because the system punishes women who refuse the performance.

There is also an economic layer. When women's work is tied to a platform identity, the platform becomes a gatekeeper for income. A creator can lose monetization, lose reach, or lose sponsorship opportunities overnight because the platform decides her content is risky. Even women who are not "influencers" feel this pressure. Employers look at profiles. Clients expect online presence. Institutions treat a platform identity as a credential. That makes women dependent on systems that can withdraw legitimacy without appeal.

And then there is the problem of permanence. Centralized platforms keep records even when women want closure. Old posts surface. Old photos resurface. Old relationships become searchable again. A woman can outgrow a phase of her life and still be judged by its digital residue. At the same time, centralized platforms can erase women's histories instantly when it benefits the company. So women get the worst combination: platforms remember what can be used against them and delete what women want to preserve.

This is why women feel unstable online even when nothing "bad" happens. Their identity is held in a system that can change without warning. The platform can shift political priorities. It can shift moderation policies. It can shift algorithmic incentives. It can shift what it considers acceptable. And because the platform owns the identity layer, women have no leverage. They can comply, perform, or disappear.

The constraint is not simply censorship. It is identity ownership. When identity is owned by an institution, speech becomes conditional. Community becomes conditional. Work becomes conditional. Safety becomes conditional. Women do not lose power all at once. They lose it through dependency on systems that can revoke access while claiming neutrality. And as long as women believe follower counts equal control, they will keep investing in identities they do not actually own.

The Shift

Once women understand that platform identity is rented, the goal becomes clearer. The goal is not to win status inside a system that can revoke you. The goal is to hold an identity that cannot be taken away by policy, pressure, or corporate mood. When identity is portable, everything else changes. You can speak with less fear. You can build community with less fragility. You can create work without wondering if a platform will erase it.

This is what Nostr changes at the structural level. It replaces platform-owned identity with user-held identity. Instead of creating a profile inside a company's database, you create a keypair. Your public key is your identity in the network. Your private key is your control over that identity. No platform can reset it for you, and no platform can take it away from you. That sounds intimidating at first, but the real concept is simple: you own the account because you hold the key.

Most women have been trained to think of identity as something confirmed by outside systems. Verification badges. Official profiles. Real-name rules. Linked phone numbers. Centralized recovery. Those systems are framed as safety, but they also create dependency. They turn identity into a permission relationship. You are "real" because an institution says you are real. And once identity is built this way, compliance becomes part of staying visible. Women learn to behave in ways that keep them legible, predictable, and acceptable.

Nostr shifts that dynamic because identity does not require institutional approval. Your identity exists because you created it, not because you passed a trust pipeline. That means a woman can build a durable public voice without tying it to a government document, a phone number, or a platform's verification process. This is not about hiding. It is about refusing unnecessary exposure. It is about building a presence that is harder to pressure through identity linkage.

When women control identity, follower counts shrink in importance. The numbers stop being the point. A following on a centralized platform feels valuable because it cannot travel with you. You cling to it because losing it feels like losing your voice. On Nostr, your identity is not trapped inside one place, so community becomes more durable. If you switch clients, you keep your identity. If one relay fails, you can use another. If a specific interface becomes hostile or unusable, you can move without starting over. That changes your bargaining position.

This is the shift: identity becomes infrastructure, not reputation. Reputation still matters, but it is built on top of an identity you own. It is not owned by a company that can rewrite it. A centralized platform can make you look "untrustworthy" by throttling your reach or restricting your account. It can reshape your public face through silent enforcement. On Nostr, no single company can do that to your identity. People may disagree with you socially, but the network does not have a central switch for legitimacy.

The shift also changes how women handle privacy. Centralized platforms push women toward a single unified profile because unified profiles create better data. Better data means better targeting, better surveillance, better monetization. Nostr supports separation because keys can be separated. A woman can choose to have one identity for public writing, another for private community, another for professional work, another for family communication. She can keep boundaries without needing permission. This is not paranoia. It is a rational response to an environment where women pay higher penalties for exposure.

Controlling identity also changes how women handle mistakes and growth. On centralized platforms, identity is sticky in the worst ways. Old posts can be resurfaced and used as evidence. Old beliefs can be used to define you permanently. At the same time, the platform can erase your history instantly if it chooses. Women get judged by what remains and punished by what disappears. With a portable identity, women can keep continuity where it matters and draw lines where it is needed. They can build archives intentionally. They can publish without being trapped by a company's timeline logic.

This shift does not mean women should abandon every mainstream platform. Most women will still use centralized systems for reach, discovery, and convenience. The shift is not withdrawal. The shift is redundancy. Women build their public identity in more than one place so no single institution owns their social existence. Nostr becomes the continuity layer. It becomes the place where your identity is stable, even if everything else becomes unstable. It becomes a channel that can survive changing corporate rules and shifting political climates.

It also changes how women relate to safety. Many women think safety comes from stronger enforcement and more centralized protection. In practice, those systems often fail women and still demand exposure as the price of participation. A platform may require identity verification but still allow harassment. It may offer reporting tools but still amplify outrage. Women learn that "trust and safety" language does not guarantee safety outcomes. When women hold identity directly, safety becomes more grounded. Not perfect, but grounded. You can reduce the amount of leverage strangers and institutions have over your life.

The shift is also psychological. Women have been trained to believe that legitimacy comes from being recognized by a platform. That is why badges matter. That is why follower counts feel like proof. That is why losing an account feels like being erased. Nostr interrupts that training. It gives women a way to be present without being granted permission. It gives women a way to build credibility through consistency rather than platform endorsement. It gives women a way to speak without treating the platform as the judge.

This is where the topic becomes larger than social media. Identity is not only about posting. It is about access to networks, access to opportunities, access to resources. In high-control environments, identity becomes a tool of governance. When identity is tied to centralized systems, it becomes easier to monitor, restrict, and pressure. When identity is portable, pressure becomes harder to apply cleanly. The point is to reduce how easily you can be managed through identity dependency.

As constraints accumulate, this shift becomes essential. Control increases through existing systems long before new rules are announced. Identity systems become stricter. Verification becomes mandatory. Speech becomes traceable. Pressure becomes quiet and procedural. Women who treat identity as infrastructure are less likely to be caught off guard. They are not immune to consequences, but they are less disoriented by them. Disorientation is one of the most effective tools of control. The shift does not promise safety. No identity system can. What it offers is clarity about where power sits and how much of your life you are handing over.

This chapter does not ask women to chase visibility. It asks them to own identity. It explains why follower counts are not control, why platform verification is not protection, and why systems like Nostr matter because identity becomes portable instead of permission-based.

Identity stops being a profile you rent and becomes what it has always been. A foundation. One that can support speech, community, and work without requiring constant approval. The shift is not ideological. It is practical. When women understand that legitimacy does not have to be granted by a platform, they stop building their lives on borrowed access. They begin to build identities they can carry across changing conditions. That decision is the beginning of autonomy.

~6
Speech Across Borders

The Constraint

Women are often told that censorship is something that happens "somewhere else." In other countries. Under other regimes. In places where the government is openly brutal and the rules are clearly enforced. In the United States and similar Western systems, speech is framed as stable. Protected. A cultural value. A constitutional guarantee. But global communication reveals a different reality. Speech is never just a right. It is always a system. And systems behave differently depending on power, pressure, and incentives.

The first constraint is that women live inside multiple speech environments at once. A woman can post something in her home country and be punished by people in another. She can speak to her own community and still be moderated by a platform headquartered elsewhere, following policies shaped by foreign laws, corporate reputations, and international lobbying pressures. Global platforms are not neutral bridges between countries. They are private governance structures that operate across borders with their own priorities.

This matters because what counts as "acceptable speech" is not universal. In some places, criticizing the government is a crime. In other places, criticizing certain social categories is treated as violence. In some places, religious offense is punishable. In other places, political misinformation is the primary target. In some places, sexual speech is heavily restricted. In others, it is commodified and encouraged. Platforms navigate this by doing what corporations always do: they minimize risk and protect access to markets. Women are caught inside that calculation.

The second constraint is that platforms enforce rules globally even when speech is local. A woman's post may be legal in her country and still be restricted because the platform is complying with regulations elsewhere. The user experiences this as confusion. She did nothing illegal. She violated no local norm. Yet her speech is suppressed, her account is limited, or her content is labeled harmful. This is not a glitch. It is the logic of centralized governance. The platform applies one rule set, or a shifting patchwork of rule sets, based on its own incentives not based on the user's reality.

The third constraint is that global speech systems create asymmetry. Some people can speak freely because they have institutional protection. Others cannot. Women, especially women without wealth or status, often carry higher risk simply for being identifiable. When speech is tied to identity and identity is tied to location, the consequences of being visible change dramatically. A woman in one country may be criticized online and move on with her day. A woman in another country may be arrested, interrogated, or harmed. The internet makes these realities collide while pretending they are the same.

This is why women cannot treat speech as purely symbolic. Speech is operational. It affects employment, legal exposure, social standing, family safety, and physical security. When platforms treat speech as content and women treat speech as survival, there is a mismatch. The platform's priority is global stability. The woman's priority is local risk. And global stability often comes at the expense of women who are easiest to sacrifice.

The fourth constraint is surveillance. Global communication runs through systems designed to collect data. Even when censorship is not visible, monitoring is constant. Messages are logged. Metadata is stored. Devices are tracked. Accounts are linked. In many countries, surveillance is not a theory. It is a fact of life. But even in countries where surveillance is framed as "security," the effect is similar: women are watched, categorized, and evaluated by systems they cannot see.

Women often assume surveillance targets "bad actors." In practice, surveillance targets patterns. Networks. Communities. Movements. Anyone who becomes noticeable. This is why women's speech across borders is risky. A woman can be harmless and still be flagged because she connects to the wrong topics, the wrong people, or the wrong timing. The system does not need to be personally interested in her. It only needs to treat her as part of a cluster.

The fifth constraint is that global platforms cooperate with states, sometimes openly and sometimes quietly. They share data. They comply with takedown requests.

They adjust moderation settings by region. They modify what can be searched, what can be recommended, and what can be monetized. A woman might believe she is speaking freely because her post is still visible to her. Meanwhile, the platform may be limiting reach in specific regions, suppressing distribution, or removing discoverability. Control can be local even when the platform is global.

Women also face the constraint of linguistic vulnerability. Automated moderation works worse across languages and dialects. Sarcasm, slang, reclaimed words, and cultural context are often misread. A woman's post can be flagged because the system does not understand her language. Or because it understands it too literally. Or because it is trained on datasets that treat certain speech patterns as inherently risky. Women who speak in non-dominant languages are often punished more because the system is less accurate and more aggressive.

Then there is the issue of exportable punishment. When a woman becomes visible across borders, her speech can be reported to her employer, her school, her family, her government, or her community. Digital speech is portable in the worst ways. A comment made in one context becomes evidence in another. A woman's identity becomes searchable across systems. Her words are detached from her intent and reattached to whatever narrative benefits the person weaponizing them. Global communication expands reach, but it also expands the range of consequences.

This is why women experience the internet as unstable. The rules are not consistent. The enforcement is not transparent. The punishment is not proportional. And the stakes are not evenly distributed. A man can treat online speech as debate. A woman often has to treat it as a risk map. She has to ask what happens if the wrong person sees this. What happens if it is screenshotted. What happens if it reaches a different country's norms, laws, or enforcement culture.

Global communication reveals that censorship is not only a government issue. It is a power issue. Corporations govern speech. States pressure corporations. Automated systems amplify enforcement. And women are positioned at the intersection, carrying the costs because they are less protected and more exposed. The constraint is that speech across borders is not simply speech. It is jurisdiction, surveillance, platform incentives, and unequal consequences, all layered on top of a single post.

The Shift

Once you understand that speech is not experienced equally across borders, the goal changes. The goal is not to "speak freely" in some abstract sense. The goal is to build communication that can survive uneven rules, shifting enforcement, and unequal consequences. Women do not need fantasy freedom. They need functional resilience.

The first shift is to stop treating platforms as neutral translators between countries. Global platforms are not public utilities. They are corporations managing risk across markets. That means their speech policies will always tilt toward whatever keeps them profitable and protected. Women who understand this stop expecting consistency. They stop expecting fairness. They stop expecting the platform to prioritize their safety when compliance with regulators and advertisers is on the line.

The next shift is to treat speech as layered. There is what you say, but there is also where you say it, who can see it, how long it lasts, and what metadata is attached to it. Across borders, those layers determine risk more than the words themselves. A harmless statement in one place can become criminal speech in another. A political opinion in one country can become a surveillance trigger in another. The shift is to understand that speech is an environment, not just content.

This is where decentralized infrastructure matters. Nostr does not erase borders, but it changes the choke points. It reduces how much global speech depends on a single company deciding what is "allowed" everywhere. When one platform controls the identity layer, the feed layer, and the distribution layer, cross-border pressure becomes easy. A government pressures the company, the company complies, and millions of people are shaped without being told. Nostr reduces that leverage by spreading the network across relays and clients, and by making identity portable rather than platform-owned.

That portability matters more than most women realize. Across borders, identity is often the vulnerability. When speech is tied tightly to a real-world identity, consequences become easier to impose. Employers can punish. Institutions can retaliate. Governments can monitor. Families can pressure. Nostr allows women to control identity more deliberately. That does not mean anonymity is automatic, but it means women are not forced into one legible profile that everything attaches to. Women can choose when to be public, when to be private, and when to separate contexts.

The shift also means accepting that safety is not granted by rules. Many women assume that stronger platform policies will protect them. But in global contexts, policy is often just a mask for compliance. Platforms will restrict speech to appease regulators. They will share data to maintain market access. They will adjust enforcement in ways users never see. Women cannot build cross-border resilience by trusting policies written for corporate protection. They build resilience through structure: redundancy, portability, and controlled exposure.

This is why decentralized communication becomes practical, not ideological. When a platform is centralized, it becomes a single global governor. It can silence a woman in one country because of pressures in another. It can restrict topics globally because they are politically costly somewhere. It can tune algorithmic distribution to match state preferences without ever announcing it. Nostr does not prevent all suppression, but it makes suppression fragmented rather than total. A relay can refuse content, but other relays can carry it. A client can filter something, but other clients can display it. That is what makes censorship harder to enforce cleanly.

The shift is also about building a stable contact point. Across borders, women often lose access suddenly. Accounts get locked. SIM cards get restricted. Phone numbers change. Apps get removed from stores. Links get blocked. A woman's network disappears because a centralized system collapses the moment one credential fails. Nostr gives women a continuity layer that can survive those disruptions. Your identity is your key. Your connections can be rebuilt across clients. Your voice does not have to die because one platform decides you are inconvenient.

Women also need to understand metadata as part of speech. Surveillance is not only about reading posts. It is about tracking who speaks to whom, when, and how often. It is about mapping networks. In cross-border contexts, metadata can matter more than content. The shift is to reduce unnecessary linkage. Women can separate identities. They can avoid attaching location and routine to public speech. They can stop posting in ways that make their offline life easy to map. This is not fear. It is strategic restraint.

Nostr supports that restraint because it does not demand a single unified profile tied to a phone number and a government-backed identity. Women can choose clients and practices that reduce exposure. They can treat online identity the way they treat physical safety: with boundaries, not with denial. This shifts women from being legible by default to being deliberate by design.

The shift also reframes what "global community" means. On centralized platforms, global connection is real but fragile. It is owned by a company that can redraw boundaries overnight. It can cut off regions. It can block links. It can suppress content to keep market access. Nostr does not guarantee harmony, but it makes communication less dependent on one institution's permission structure. That matters when women are trying to share information across borders, build parallel networks, or maintain contact through political turbulence.

This is not an argument for constant political posting. It is an argument for durable channels. Women do not need to fight every censorship battle online. They need to avoid being trapped when the battle comes to them. The shift is to build redundancy before you need it. That means learning a system like Nostr while things are still functional. It means securing your identity before enforcement becomes strict. It means building a network before you are isolated.

As power concentrates, this becomes essential. Control intensifies through existing systems long before new rules are announced. Platforms become enforcement partners. Speech becomes traceable. Women who understand cross-border dynamics are less likely to be caught off guard. They are not immune to pressure, but they are less disoriented by it. Disorientation is one of the most effective tools of control. The shift does not promise safety. No communication system can. What it offers is continuity: continuity of identity, continuity of connection, continuity of access when centralized channels fail.

This chapter does not ask women to treat speech as symbolic. It asks them to treat it as structural. It explains what global communication reveals about censorship and power, and why decentralized infrastructure like Nostr gives women more options when rules, enforcement, and consequences shift across borders.

Speech stops being a privilege granted by a platform and becomes what it has always been. A capacity. One that can be protected, diversified, and carried across changing environments. The shift is not ideological. It is practical. When women understand that borders change the cost of speech, they stop assuming one set of rules will protect them. They begin to build communication systems that survive uneven power. That decision is the beginning of resilience.

~7
Community Without Gatekeepers

The Constraint

Most women have been taught that online community is something you join, not something you shape. You download the app, you create the account, and you enter the space the platform gives you. The platform decides what kind of community is possible. It decides what gets recommended, what gets buried, what gets labeled "harmful," and what gets boosted as "healthy conversation." Women are told this is normal. They are told community is just culture. In reality, community is infrastructure. And infrastructure always has gatekeepers.

The first constraint is that gatekeepers are often invisible. Most women are not blocked by a person standing at a door. They are blocked by systems. Search results that do not show them. Posts that do not travel. Comments that disappear. Accounts that get quietly throttled. Communities that get "restricted" or "limited" for safety. The platform rarely has to say no directly. It only has to make participation harder until women stop showing up.

This changes who gets heard. Centralized platforms reward voices that fit their incentives. They reward content that keeps people scrolling. They reward conflict because conflict generates engagement. They reward soft compliance because soft compliance keeps advertisers comfortable. Women who speak plainly about power, coercion, sexuality, money, abuse, politics, religion, or violence are often punished without explanation. The platform does not have to disagree with them. It only has to treat them as risky.

The second constraint is that centralized communities do not belong to the members. They belong to the platform. A group may feel like a home. It may have shared language, shared care, and shared history. But the infrastructure is owned by someone else. A policy update can erase it. A moderation decision can fragment it. A rumor can trigger mass reports. A cultural shift can mark the entire community as "problematic." When that happens, the people inside the group learn the hard lesson: community inside a platform is always conditional.

Women experience this more sharply because women's communities are often built around survival. Mutual aid. Safety information. Warning systems. Emotional support. Exit strategies. Knowledge about work, family, medical systems, and abuse. These are not casual interests. They are sometimes the difference between coping and collapse. When that kind of community is disrupted, women do not just lose entertainment. They lose structure.

The third constraint is that gatekeeping is often outsourced to mobs. A platform can claim neutrality while letting users punish each other. Dogpiles become enforcement. Quote-tweets become punishment rituals. Reporting tools become weapons. Visibility becomes a threat. If a woman is targeted, she is expected to absorb the impact quietly. The platform may claim it cannot intervene. Or it may intervene by punishing the wrong person, because automated systems cannot tell the difference between harm and defense.

This creates a predictable behavior pattern. Women learn to monitor themselves. They soften language. They add disclaimers. They avoid saying names. They avoid direct criticism. They avoid organizing publicly. They stop posting questions that might get mocked. They stop telling the truth in a way that would actually help other women. Not because they are weak, but because the cost is real. A centralized platform can convert one honest post into a full-scale social penalty.

The fourth constraint is that centralized communities are shaped by algorithmic incentives. The platform decides who is connected to whom. It decides which posts are seen first. It decides which arguments get amplified. It decides what kind of tone travels. And because the algorithm is tuned for engagement, it favors communities that produce constant reaction. That means the healthiest communities are often the least visible. They are not profitable enough to be featured. Meanwhile, the most unstable environments get rewarded because they keep people emotionally hooked.

Women also face a constraint of pace. Centralized platforms push everything toward speed. Faster replies. Faster outrage. Faster hot takes. Faster content churn. But real community requires time. It requires context. It requires memory. It requires patience. Women are often carrying work, caregiving, stress, and real danger in their lives. They do not have endless time to perform constant engagement to remain "seen." The platform interprets that as low relevance. In practice, it means women's communities are punished for being human.

The fifth constraint is that centralized platforms collapse boundaries. They merge personal and public space. A woman can be trying to talk to a small group of women about safety, and the content can still be pulled into a public arena. Her audience becomes a crowd. Strangers arrive. The tone shifts. Harassment enters. The platform calls that "reach." The woman experiences it as exposure. This is how women lose the ability to speak honestly, because honesty becomes a liability when the room cannot be controlled.

Gatekeepers also control identity. Most platforms require legibility. Phone numbers. linked accounts. verification. consistent profiles. That means communities become easier to map. Easier to monitor. Easier to infiltrate. Easier to punish. When a woman's speech is tied to a platform-owned identity, community becomes another place where her life can be made vulnerable. She is not just participating. She is leaving a trail.

And because the platform owns the community layer, women are always working under the threat of removal. A group can be deleted. A moderator can be banned. A shared archive can vanish. A key account can be limited. A network can be fragmented without warning. This fragility teaches compliance. Women learn to avoid saying what would trigger enforcement. They learn to keep community within a narrow range of acceptable conversation. They learn to behave as if the gatekeepers are always listening.

The constraint, then, is not that women cannot find community online. The constraint is that community inside centralized systems is always shaped by someone else's power. The gatekeepers set the boundaries. The algorithm shapes the culture. The platform owns the continuity. Women can build relationships inside it, but they cannot secure the structure. And when women cannot secure the structure, they cannot fully trust the community to remain. That uncertainty becomes part of the cost of belonging.

The Shift

Once women understand that centralized communities are conditional, the goal changes. The goal is not to find a perfect platform with perfect rules. The goal is to build community that cannot be erased by a single policy update, a single moderator decision, or a single corporate panic. Women do not need louder spaces. They need durable ones.

The shift begins with a different definition of "gatekeeper." Most women imagine gatekeepers as people. Moderators. Influencers. Admins. But on centralized platforms, the real gatekeeper is infrastructure. Algorithms decide who is discoverable. Search decides who is findable. Recommendation systems decide what becomes visible. Monetization systems decide what gets protected. Enforcement systems decide what gets punished. Women can behave perfectly and still be buried because they are not profitable, not legible, or not convenient.

Nostr changes this by reducing the number of gates that can close at once. It does not create a single community space owned by one entity. It creates a protocol layer where many communities can exist without being dependent on one platform owner. That means no single company can delete the entire environment. A specific client can choose what to show. A relay can choose what to carry. A community can choose what it tolerates. But the woman's identity and ability to publish are not tied to one central authority.

It does not offer comfort. It offers options. Options matter because options prevent total collapse. On centralized platforms, women lose communities all at once. The group disappears. The archive disappears. The key accounts disappear. The shared language disappears. The exit routes are unclear because the platform never wanted exit routes to exist. On Nostr, fragmentation is built into the structure. That can feel messy, but it also prevents one decision from wiping out everything.

The shift also changes who gets heard. On centralized platforms, voices rise through algorithmic incentives. Content that triggers outrage travels farther. Content that creates conflict gets rewarded. Communities become shaped by whatever generates reaction, which means the loudest voices often dominate even when they are not useful. Nostr can still have noise, but it is less dependent on a single engine that amplifies conflict at scale. Visibility becomes more social than algorithmic. You are heard because people choose you, not because the machine decided you should trend.

This matters for women because many women are not trying to win attention games. They are trying to find signal. They are trying to locate trustworthy people, steady information, and communities that do not punish them for being direct. Centralized platforms do not optimize for that. They optimize for addiction, churn, and engagement. The shift is moving into spaces where the community structure can be built around values instead of profitability.

Nostr also shifts the relationship between community and identity. In centralized systems, identity is often forced into one permanent profile. That means every community you join becomes attached to a single legible record. It becomes easy to map where you spend time, what you believe, and who you talk to. Women lose privacy through participation. Nostr makes separation possible. Women can choose different keys for different roles and different contexts. They can build community without automatically giving strangers a full map of their life.

This is where women regain control of boundaries. In centralized systems, boundaries are treated as hostility. A woman who blocks people is accused of being fragile. A woman who refuses debate is accused of being weak. A woman who protects her privacy is accused of hiding something. Those accusations are cultural tools used to push women back into exposure. Nostr supports a different posture. Women can curate aggressively without having to justify it to a platform's engagement incentives. They can stop treating community as something that must remain open to everyone.

The shift also changes how harassment behaves. Most harassment works best when there is a shared stage and a shared crowd. One quote-tweet can summon thousands. One viral clip can turn a woman into a target. Centralized platforms are designed to produce that aggregation. Nostr does not eliminate hostility, but it reduces the automatic mass amplification that turns one moment into a global dogpile. It becomes harder to manufacture a mob when there is no single algorithm pushing the spectacle into everyone's feed.

That does not mean women should expect a harassment-free environment. They should expect reality. The difference is that women can build systems where harassment is easier to route around. They can change clients. They can change relays. They can curate networks. They can filter aggressively. They can build smaller communities that do not depend on mass visibility. The goal is not perfection. The goal is recoverability.

Community also becomes more durable when it is treated like infrastructure instead of entertainment. Most women have been trained to treat social platforms as social life. The platform becomes the container for friendship, politics, work, news, and identity. That creates dependency. If the container breaks, everything breaks. The shift is to build community as a layered system. Nostr can be one layer. Email can be another. Private chats can be another. Real-world relationships can be another. Women are safer when community is diversified.

This shift is not dramatic. It is quiet. It looks like building a reliable contact point that is not controlled by a platform owner. It looks like having a place where people can find you even if a mainstream account disappears. It looks like building trust slowly instead of chasing scale. It looks like choosing smaller communities with stronger norms instead of trying to survive in the most chaotic public feeds. It looks like refusing the idea that community must always be public to be real.

It also means women stop confusing "being seen" with "being supported." Centralized platforms can make a woman visible and still leave her alone when she is targeted. They can amplify her content and still punish her for the backlash it creates. They can sell her visibility as empowerment and then treat her as a liability when she becomes inconvenient. The shift is to build community around mutual recognition, not around algorithmic exposure.

When conditions tighten, this difference becomes obvious. Communities that depend on a platform owner become fragile fast. They can be throttled or removed with no warning. Women who built everything inside one system are easier to isolate, because isolation becomes an administrative action. Women who built parallel channels are harder to cut off completely.

This chapter does not ask women to find perfect communities. It asks them to stop building community under gatekeepers who can erase it. It explains how decentralized spaces change who gets heard, how attention behaves, and why a protocol like Nostr gives women more options when platforms become unstable.

Community stops being a room you borrow and becomes what it has always been. A network. One that can be built, protected, and carried forward. The shift is not ideological. It is practical. When women understand that community is never neutral online, they stop expecting platforms to preserve what matters. They begin to build relationships and structures that survive pressure. That decision is the beginning of durability.

~8
Risk, Harassment, and Reality

The Constraint

Decentralization is often described as relief. No corporate gatekeeper. No central moderation team controlling what is allowed. No algorithmic manager silently shaping behavior. For women who have been punished or erased by centralized systems, that promise can feel like a way out. But decentralization is not a guarantee of safety, and it is not a guarantee of kindness. It changes the structure of control. It does not change human behavior, and it does not remove the incentives that produce harassment.

The first constraint is that decentralization shifts responsibility onto the user. Centralized platforms are hostile and inconsistent, but they still provide certain built-in functions: account recovery, unified blocking tools, and one place to report abuse. Those systems often fail women, but they exist as defaults. In decentralized spaces, the environment is not one coherent "platform." It is a network of clients, relays, and communities, each with different norms and different limitations. A woman is not entering a single room. She is navigating a moving landscape.

This matters because women do not experience risk as abstract. Risk is logistical. It shows up as interruptions, unwanted attention, exhaustion, and time loss. If a woman joins a chaotic relay or uses a weak client, she may be flooded with noise, bait, pornography, spam, or aggressive speech designed to provoke. None of this is "illegal," so the system will not automatically stop it. The question becomes whether the woman has the tools and patience to filter it. The cost is not only emotional. It is operational.

The second constraint is that some people interpret decentralization as permission to dominate. They confuse "no central authority" with "no consequences." They treat women as targets or as entertainment. They test boundaries to see what they can get away with. They call harassment "honesty," call threats "jokes," and call disrespect "free speech." This is not unique to decentralized networks, but it becomes more visible when there is no corporate manager smoothing the surface. Women who arrive expecting a calmer world can be shocked by the bluntness.

There is also the reality of culture lag. Many decentralized communities are young, fragmented, and unevenly moderated. Some have strong norms and skilled builders. Others are held together by loud personalities and shallow status games. Women can walk into a space where people act as if cruelty is a sign of strength. If she objects, she may be told she is too sensitive or doesn't understand the culture. That dynamic is familiar. The structure may be different, but the social pattern repeats.

The third constraint is that decentralization does not prevent coordinated attacks. It reduces certain forms of mass amplification, but it does not remove hostility. People still screenshot. People still quote. People still misrepresent. People still build narratives around a target and circulate them. In decentralized networks, the dogpile can be more distributed and harder to track, which can make it feel more chaotic. A woman might not even know where the hostility is coming from. She just feels the pressure arrive.

Women also face a constraint of social engineering. In centralized platforms, the obvious risk is platform punishment. In decentralized spaces, the risk often comes from other people using access to extract information. Friendly messages can be probes. Compliments can be bait. Invitations can be traps. A woman may be nudged to reveal identifying details, share private keys, click unsafe links, or join off-platform chats where boundaries are weaker. The attack is not always loud. It can look polite until it becomes controlling.

The fourth constraint is that decentralization does not eliminate hierarchy. Many women assume that removing the platform owner removes power dynamics. It does not. Influence still forms. Status still forms. Social capital still forms. Some accounts become central. Some communities become dominant. Informal leaders emerge and shape what is acceptable inside their circles. If women treat decentralization as automatically egalitarian, they will miss the new forms of gatekeeping as they form in real time.

There is also the constraint of inconsistent safety tools. Some Nostr clients have strong mute lists, filters, and reporting workflows. Others are minimal. Some make it easy to block harassment. Others make it inconvenient.

Some make it easy to separate identities. Others push women into default exposure. In centralized platforms, women are harmed by the platform's incentives. In decentralized spaces, women can also be harmed by weak tools and careless design. A network can be decentralized and still be unfriendly to women.

Another constraint is the myth of anonymity. Decentralization is not invisibility. Posts can be public. Messages can be saved. Screenshots can be archived. Writing styles can be recognized. Patterns can be mapped. Even without a central company collecting everything in one place, the internet still remembers. Women who assume decentralization equals anonymity often take risks they would not take elsewhere. The danger comes from false confidence, not from the protocol itself.

There is also permanence without accountability. In centralized platforms, content can be removed quickly when a company decides it is risky. In decentralized spaces, content can persist longer, especially if it spreads across relays and clients. That can protect women from unfair erasure. It can also lock women into posts they wish they could take back. A moment of oversharing, grief, or anger can remain accessible longer than she intended. Women deserve space to be real, but they also deserve the ability to retreat without a permanent trail.

The fifth constraint is that women are often expected to do the safety labor themselves. If a space is harsh, the implicit message is that women should filter harder, block more, and adapt faster. That expectation mirrors the broader world. Women are told to manage risk privately instead of expecting environments to be designed responsibly. Decentralization does not automatically change that. It can even intensify it, because there is no single authority to demand better behavior from. Women must choose their tools and communities carefully, or they pay the price.

The constraint is not that decentralized networks are bad. The constraint is that decentralization does not fix human nature, and it does not remove the social penalties women face online. It removes one kind of gatekeeper and introduces a different set of responsibilities. It creates new freedoms, but it also creates new vulnerabilities. Women cannot approach decentralized spaces as utopias. They have to approach them as real environments: imperfect, uneven, and full of both opportunity and risk.

The Shift

The shift begins when women stop confusing decentralization with safety. A system can remove platform owners and still concentrate cruelty. It can eliminate central censorship and still reward harassment. It can give you control over identity and still give other people endless access to your attention. Nostr changes the architecture of speech, but it does not change human behavior. The risk moves. It does not disappear. The first advantage is clarity. The first protection is realism.

Most mainstream platforms keep order by enforcing rules from the top. That enforcement is inconsistent and often weaponized, but it still exists as a friction layer. Nostr reduces that centralized leverage, which matters when speech systems tighten and permission becomes a tool of governance. But when you remove a centralized gatekeeper, you also remove centralized containment. There is less institutional protection. There is also less institutional denial. You are closer to the real conditions of the network. That is the trade. Many women are not prepared for it, because most people were trained to treat "community guidelines" as safety rather than as corporate risk management.

The shift is understanding that moderation is not the same thing as accountability. Centralized platforms are often abusive in how they enforce rules, but they can still remove a stalker's account or block a threat cluster quickly when it becomes inconvenient enough. Decentralized networks do not have a single authority with that reach. You can block people. You can leave relays. You can change clients. But there is no one to appeal to, and no one to blame. That can feel like freedom until a woman is the target. Then it can feel like being alone in an open field.

This is where women need a different definition of "safe." Safe does not mean no hostility. Safe means you can limit exposure, preserve your energy, and keep your identity from being trapped inside the worst moment of attention. On Nostr, you get more control over continuity. You can keep your identity across clients. You can choose which relays carry your posts. You can exit one interface without losing everything. But you also have to take responsibility for what you see, where you post, and how you respond. The shift is treating that responsibility as a strength, not a burden.

Harassment on decentralized systems often looks more direct because there is less performative compliance. On mainstream platforms, attackers often operate through mass reporting, dogpiles, quote-tweets, and indirect humiliation. On Nostr, it can be simpler. Someone replies. Someone tags you. Someone tries to bait you into attention. The design does not guarantee kindness. It only reduces centralized power. That means the woman's real defense becomes her boundaries, not the platform's promises.

For women, the most important boundary is attention. Harassment thrives on access. It thrives on forcing you to react, explain, defend, or perform emotional labor on demand. On centralized platforms, the system already extracts that labor through visibility incentives and social punishment. On Nostr, you are not required to play that game. But you may still feel compelled. The shift is recognizing that you are allowed to be unreachable. You do not owe strangers dialogue. You do not owe predators closure. You do not owe bad-faith people education. Decentralized speech is still speech. It is not a courtroom. Your refusal to engage is not weakness. It is structural control.

Another part of the shift is distinguishing identity control from reputation control. Nostr gives you stronger identity continuity, but it does not give you reputation immunity. Screenshots still exist. People still misrepresent you. Narratives still travel. The difference is that your identity is harder to erase through a single corporate decision.

That matters in environments where enforcement expands through routine mechanisms and speech becomes easier to punish through existing systems. But social punishment still functions. Women still face tone-policing, sexualization, and targeted hostility. Nostr does not remove misogyny. It removes some of the centralized choke points that make removal automatic.

This shift also clarifies that decentralization changes what "community" means. There is no single Nostr community. There are clusters, norms, and subcultures. Some are thoughtful and disciplined. Some are chaotic. Some are hostile. A woman can find genuine support on Nostr, but she can also find relentless conflict. The network will not curate her world for her. That can be a gift, but only if she stops expecting comfort as the default. Decentralized spaces require selection. The woman chooses who she follows, what she tolerates, and what she exits. The shift is deciding that her time and nervous system matter more than proving a point to strangers.

There is also a reality women should accept early. Bad actors use open systems too. The fact that a system is decentralized does not mean it is morally better. It means control is distributed. That can protect women from centralized erasure, but it can also create persistent exposure to people who enjoy being cruel.

The advantage is that a woman can respond structurally instead of emotionally. She can build habits of containment. She can treat harassment as noise. She can mute, block, and move on without treating every threat as a personal referendum. That is the kind of emotional discipline modern speech systems try to steal from women.

The shift is not to become paranoid. It is to become hard to destabilize. Women do not need to be online without fear. They need to be online without dependency. Dependency is what makes harassment effective. Dependency on a single platform, a single account, a single audience, a single social identity. Nostr reduces that dependency by giving women portability. But portability only matters if a woman is willing to leave hostile spaces without hesitation. The healthiest relationship to Nostr is not loyalty. It is optionality.

This chapter explained why decentralization does not automatically produce safety or kindness. It showed how harassment, bad faith engagement, and exposure still exist even when platform owners disappear. The shift is learning to treat Nostr as infrastructure, not as protection, and to treat boundaries as a structural tool rather than as a personal weakness.

The final reality is simple. Speech without permission does not mean speech without consequence. It means consequence is no longer controlled by one company's policy department. That can be a relief in hardening environments where enforcement expands through the systems already in place. But it also means women cannot outsource safety to rules or moderation theater. Nostr gives women more control over identity and continuity. The rest is choice. What she tolerates. What she ignores. What she refuses. That is not a downside. That is the point.

~9
When Nostr Makes Sense

The Constraint

Nostr makes sense when you understand the constraint it responds to. The constraint is not that mainstream platforms are imperfect. It is that they are permissioned systems. A woman can build years of work, relationships, and visibility inside an app and still lose it overnight through a policy change, a lockout, a moderation decision, or a quiet distribution shift. The platform may not need to ban her to control her. It can simply make her harder to find, harder to reach, and easier to ignore. That is what dependency looks like in a networked world.

This matters because women are often building under pressure. She is running a household, maintaining relationships, working, caregiving, publishing, or trying to leave a situation that is unsafe. The cost of disruption is not abstract. When access to communication is interrupted, a woman loses time, coordination, income, and sometimes safety. She cannot treat platforms as entertainment when they are also infrastructure. She needs channels that still work when the environment tightens, when enforcement becomes automated, or when attention systems become less forgiving.

The first constraint is identity ownership. Corporate platforms issue identity the same way banks issue access. You exist there because a company says you can. Your username, your history, and your follower graph are not portable. They are assets inside a container. If the container changes its rules, you adjust or you disappear. That is why centralized platforms feel stable until they do not. They work smoothly for years, and then one update reveals that the relationship was never mutual.

Even when nothing dramatic happens, identity control still shows up. A platform can change what content it promotes. It can tighten link restrictions. It can downgrade posts about money, politics, sexuality, parenting, or anything that advertisers do not like. Women feel the result as silence and confusion. The feedback loop becomes distorted. She posts something true and it vanishes into the feed like it never existed. Over time she starts editing herself in advance, not because she is weak, but because she is adapting to pressure.

The second constraint is distribution control. Most platforms pretend you are publishing to your audience. In reality, you are publishing into a sorting system. The algorithm decides who sees what, when, and how often. The platform is not neutral about that decision. It is designed to keep people scrolling and reacting. That means it rewards heat over substance. It rewards conflict over clarity. It rewards whatever keeps nervous systems activated. For women trying to build calm, durable work, this turns publishing into labor that never ends.

This is where Nostr begins to make sense. It does not remove the social reality of attention, but it reduces the centralized lever. There is no single company that owns the feed logic across all clients. There is no universal choke point that can quietly rewrite distribution norms for everyone at once. Nostr, as of yet, offer more reach. It is a reduction of dependency. You are not as trapped inside someone else's visibility machine.

The third constraint is enforcement without recourse. Centralized systems lean on automation, risk scoring, and compliance frameworks. Appeals are often limited or unclear. Mistakes are common, and accountability is diffuse. When a woman is punished by an algorithm, there is rarely a human conversation waiting for her on the other side. The system records irregularity without recording context. That is the pattern. Systems built for scale remember everything and revisit nothing.

Women run into this constraint in predictable places. They talk about domestic violence and get flagged. They talk about reproductive health and get throttled. They talk about money and get accused of scams. They set boundaries and get labeled aggressive. They criticize power and get swarmed. The platform does not have to "target women" to produce gendered outcomes. It only has to enforce rules bluntly and reward mass reporting, outrage, and misinterpretation. When the burden of explanation lands on her, she loses time and stability.

Nostr does not solve this by making people kinder. Decentralization is not automatically safe or generous. But it changes the control structure. Your identity is not issued by a corporation, and your ability to publish does not depend on one platform account staying in good standing. Nostr is protocol-level resilience. That means a woman can leave one client and use another. She can switch relays. She can stay connected without asking a company for reinstatement. It is not perfect protection. It is a different architecture.

Another constraint is the collapsing of communication into performance. Centralized platforms train women to treat posting as a social job. You are expected to be available, responsive, and emotionally fluent on demand. You are expected to translate nuance into bite-sized content that performs. If you post without optimizing, you feel punished. If you post too much, you feel exposed. If you do not post, you disappear. This is a trap disguised as connection, and women are often the ones forced to manage the emotional temperature of the room.

Nostr can be useful here because it allows a different posture. It supports broadcasting without begging. It allows a woman to publish without feeling like she is auditioning for approval every time she writes. The network is quieter and less performative by default. That does not mean it is peaceful. It means it is not engineered to keep you in constant emotional motion.

Nostr also makes sense when your work is long-term. Books, independent projects, community work, mutual aid, education, and anything built on steady repetition need continuity. Mainstream platforms are optimized for the opposite. They want novelty. They want reaction. They want acceleration. When your work is meant to last, you need at least one channel that does not punish you for being consistent. Nostr cannot guarantee growth, but it can hold your signal without demanding you distort it.

There are also times when Nostr does not make sense, and that must be said plainly. If your primary need is convenience, polished interfaces, and mainstream discoverability, you will feel friction. If your main income depends on traditional platforms because that is where customers already are, you do not need to abandon that channel out of principle. Most women cannot afford purity as a lifestyle. They can afford redundancy. Nostr works best as a parallel layer, not as an all-or-nothing replacement.

It may also not make sense if you are seeking a system that provides safety through centralized moderation. Nostr gives you tools to block and filter, but it does not offer a single authority you can appeal to. You are responsible for your boundaries. That can be empowering, but it can also feel heavy if you are already exhausted. Decentralized communication is useful when you can hold your center and disengage quickly. It is less useful if you need the platform to manage the environment for you.

The simplest way to think about Nostr is as insurance against platform dependency. It is not a magic alternative. It is a second channel that reduces the risk of total interruption. It gives women a way to maintain continuity when the mainstream environment becomes more restrictive, more automated, and more hostile to anything that does not fit the approved patterns. When systems tighten, the cost of dependency rises. A woman does not need panic to justify resilience. She needs realism.

The Shift

The shift begins when you stop treating Nostr like an alternative social network and start treating it like a response to a constraint. The constraint is not boredom with mainstream platforms. It is the reality that most digital communication is permissioned infrastructure. Your identity exists because a company allows it. Your reach exists because an algorithm approves it. Your work stays visible as long as the platform's incentives remain aligned with your presence. For women, that arrangement is stable right up until it is not.

Nostr rearranges who holds continuity. It does not promise a safer internet. It does not promise kinder people. It does not guarantee an audience. It changes the architecture so that identity is portable and publishing is less dependent on a single corporate container. That difference matters most when systems tighten through quiet mechanisms. Moderation becomes automated. Enforcement becomes inconsistent. Visibility becomes conditional. The shift is realizing that these are not glitches. They are normal behaviors of centralized platforms built to manage risk and control attention.

When Nostr makes sense, it is usually because you need redundancy. Redundancy is not dramatic. It is what serious people build when they understand failure is not rare. Women already live this way in other areas of life. Backup childcare plans. Extra keys. Emergency cash. A second way home. Digital redundancy follows the same logic. Nostr is useful as a second lane that stays available when the main lane becomes unreliable.

That means the best way to use Nostr is often simple. You do not need to move your entire life there. You do not need to post constantly. You need a stable identity you control, a client you understand, and a small network of people who can still reach you when something shifts elsewhere. If you treat Nostr like a place you only visit after you have been harmed by another platform, it will feel empty and discouraging. The shift is building continuity before crisis, not after.

Nostr also makes sense when your work is long-term. Books. Publishing. Essays. Research. Organizing. Teaching. Building a project that requires months and years. Mainstream platforms are built to extract short-cycle attention. They reward conflict, novelty, and emotional volatility because those are efficient fuels for engagement. They bury calm work not because it lacks value, but because it does not create the same compulsive behavior. Nostr will not fix that by itself, but it gives long-term work a place to exist without being shaped by the same extraction logic.

This shift changes how women measure progress. On corporate platforms, metrics are treated as meaning. Likes become approval. Shares become validation. Reach becomes worth. Women are trained to adapt their voice and output to fit what performs, even when they know the system is shallow. Nostr is quieter. Feedback is smaller. The pace is less frantic. That can feel like failure if you expect the same dopamine loop. The shift is understanding that quiet is not absence.

Quiet is space. For many women, that space is what makes consistent work possible again.

Nostr makes sense when you want identity portability. In centralized systems, identity is bundled with access. If you lose the account, you lose continuity. Past work, contact networks, reader attention, reputation, and history can disappear or become unreachable overnight.

Even without a ban, distribution changes can function like a slow disappearance. Nostr changes the bundle. Your identity is not trapped inside one interface. You can switch clients and still be you. That is not a guarantee of protection. It is a reduction of fragility.

This is where the shift becomes practical. You stop asking if Nostr is "better" and start asking what it is better for. It is better for reducing single points of failure. It is better for maintaining a publishing channel that does not require algorithmic permission. It is better for building an audience relationship that is less rented. It is not better for polished convenience, mainstream discoverability, or being protected by a central authority when conflict arises. You are choosing a different set of tradeoffs.

Nostr also makes sense when you are willing to protect your attention. Decentralized systems do not manage your environment for you. They do not guarantee comfort. There is no central office to appeal to, no uniform moderation policy, no safety theater you can outsource your boundaries to. Women are often trained to respond, explain, soothe, and de-escalate. That reflex is costly in open networks. The shift is learning that you can block quickly, disengage without a speech, and preserve your nervous system without apologizing for it.

At the same time, Nostr does not make sense for every season of life. If you are already stretched thin, adding a new tool may feel like extra work instead of stability. If your income depends on mainstream platforms because that is where customers already are, you do not need to sabotage yourself for ideological reasons. Resilience is not purity. You can keep using systems that work while building redundancy slowly. The shift is choosing layered options instead of forcing an all-or-nothing identity decision.

Nostr also does not make sense if your goal is fast mass reach. You can find community and growth there, but it is not designed to push content into large audiences through aggressive discovery mechanisms. That is partly why it can feel calmer. It does not constantly amplify noise for engagement. If you need broad distribution quickly, centralized platforms will still be useful. Nostr is better understood as a parallel channel, not a replacement engine.

This is where the constraint reappears in a different form. The more centralized a system is, the easier it is to scale and smooth the user experience. The more decentralized a system is, the more responsibility returns to the individual. That responsibility includes key management, boundary setting, and choosing which clients and relays make sense for your life. The shift is not learning to do everything perfectly. It is learning enough to stay in control of continuity.

This chapter explained when Nostr is worth the time and effort, and when it is not. It showed that Nostr makes the most sense as redundancy, identity portability, and a long-term publishing channel, while centralized platforms still dominate convenience, scale, and discovery.

The shift is recognizing that is leverage. Leverage over identity. Leverage over continuity. Leverage over the ability to keep publishing when other systems tighten through automation, enforcement, and quiet suppression. For women who are building anything meant to survive pressure, leverage is not optional. It is the difference between having a voice and having a voice only as long as someone else allows it.

~10 Speaking on Your Own Terms

The Constraint

Speaking online is often framed as a personal choice. Post or don't post. Engage or ignore. Build an audience or stay private. That framing sounds empowering, but it hides the constraint underneath it. The real constraint is that modern platforms do not treat speech as speech. They treat it as data. They treat it as behavior. They treat it as a signal to be ranked, stored, and interpreted later, often without your participation.

For women, this constraint is not theoretical. Women are judged harder for being visible. They are punished faster for being direct. They are expected to explain more and tolerate more. A man can speak casually and be read as confident. A woman can speak casually and be read as unstable, aggressive, emotional, or dangerous. Visibility is not equal cost across genders. It never has been.

The first constraint is permanence. Nothing online is truly temporary anymore, even when it is presented that way. A post can be screenshotted. A reply can be clipped. A joke can become a "statement." Context is stripped first, then the record becomes the truth people argue over. Systems store everything and revisit nothing. That is a dangerous environment for anyone, and it is especially dangerous for women who are already easier to frame as untrustworthy.

The second constraint is forced performance. Centralized platforms reward constant output and constant reaction. The system prefers people who post often, respond quickly, and generate conflict or emotion. It does not reward steadiness. It rewards momentum. Over time, this turns expression into labor. You are not just speaking. You are feeding a machine that grades your presence.

For women, this becomes a trap because women are already trained to manage the temperature of social spaces. Women are expected to keep things smooth. To respond politely. To absorb tension. To "be professional" even while being provoked. Platform dynamics amplify that expectation. The more visible she becomes, the more she is expected to perform emotional regulation for strangers.

Another constraint is audience collapse. On most platforms, your audience is not one audience. It is many audiences stacked on top of each other. Friends, coworkers, strangers, critics, family members, potential customers, and people who simply want conflict are all mixed together. Speaking to everyone at once forces you to speak to no one in particular. Women often respond by shrinking their voice until it feels safe enough for the worst possible reader.

This is why silence is not always avoidance. Sometimes silence is strategy. Sometimes it is the only way to keep ownership of your nervous system and your life. The culture tells women that silence equals weakness, that not responding means you lost. That is a lie sold by people who benefit from your attention. Silence can be discipline. Silence can be power. Silence can be the boundary that keeps your identity intact.

But visibility also matters. Women need visibility for money, independence, and movement. A woman building a business, publishing books, or organizing a community cannot always stay invisible. In a tightening environment, women also need communication channels that cannot be casually taken away. The problem is that visibility has become automatic and unchosen. The constraint is that platforms default to maximum exposure because exposure is profitable for them, even when it is costly for you.

Another constraint is misinterpretation without recourse. Once speech becomes data, it becomes easy to misread. Algorithms cannot interpret tone. Strangers cannot interpret intent. People who want to harm you will interpret your words in the worst possible way on purpose. Then the system amplifies the conflict because conflict creates engagement. You are left trying to explain yourself inside a machine designed to prevent resolution. Women are especially vulnerable here because women are expected to prove innocence rather than be assumed human.

This is where "just be careful what you post" becomes useless advice. Care is not the problem. The environment is the problem. Women can be careful and still be targeted. They can be polite and still be punished. They can be quiet and still be harassed. Platforms make speech legible and searchable. They turn small moments into permanent artifacts. In systems like this, your safest move is not perfect behavior. Your safest move is deliberate control over when you speak and who you are speaking to.

There is also a constraint created by dependency on audience. When your income, identity, or sense of stability is tied to attention, attention becomes a leash. Women see this in creator culture constantly. Post even when you are sick. Share even when you are grieving. Stay consistent even when your life is breaking.

The system rewards output and punishes interruption. That makes women easy to exploit because women already carry responsibility without rest. When you depend on a platform to "keep you alive," you lose the ability to disappear when you need to.

This connects directly to decentralization. Nostr does not fix human behavior, but it changes the infrastructure around speech. It makes identity portable. It makes it easier to leave one interface without losing everything. It reduces the chance that one company can dictate the conditions of your visibility. That matters when platforms tighten quietly through enforcement, downranking, and policy shifts that always claim neutrality. A woman who has only one communication channel is easier to control than a woman with redundancy.

Still, the constraint remains: speech has a cost. Decentralized systems do not automatically lower that cost. They simply move responsibility back to the user. On open networks, you may get less safety theater and more direct consequences. There is no central authority to appeal to, and you may need stronger boundaries to stay stable. That does not mean decentralization is worse. It means you cannot outsource speech strategy to the platform. You must build it yourself.

So when does speaking on your own terms become possible? It becomes possible when you stop treating visibility as default. When you stop treating response as obligation. When you stop treating audience growth as proof of legitimacy. The constraint is that modern systems push women toward constant exposure and constant explanation. Speaking on your own terms is the act of refusing that push.

This is not about becoming invisible. It is about choosing exposure deliberately. A woman can decide when her work needs visibility and when her life needs privacy. She can decide which topics deserve public writing and which deserve silence. She can decide who gets access to her attention and who gets none. That is disciplined.

The constraint, in the end, is that platforms want your voice to be predictable, monetizable, and easy to sort. They want you posting in a way that makes you easier to rank and easier to manage. Women are pressured into this more aggressively because women are already expected to be legible and compliant. That is the foundation for everything that follows. Before tools, scripts, and tactics, a woman needs a clean understanding of the environment she is speaking into. She needs to recognize that speech is now infrastructure, and infrastructure always has a politics.

The Shift

The shift is realizing that speaking is not the same thing as being available. Most women were trained to confuse the two. If someone asks a question, you answer. If someone misunderstands, you clarify. If someone is upset, you smooth it over. Online, this becomes a trap because the environment is designed to create endless access. The more visible you become, the more people feel entitled to your time, your energy, and your explanation. Speaking on your own terms begins when you break that entitlement.

This shift is not about fear. It is about choosing where your voice actually belongs. Visibility is not automatically good. Silence is not automatically weak. Audience is not automatically your community. A woman can speak publicly and still keep herself intact, but only if she stops letting platforms decide the conditions of her presence. Most platforms default to maximum exposure because exposure is profitable. They want you discoverable, searchable, and reactive, because that produces content and conflict. The shift is choosing exposure deliberately instead of allowing it to be the default setting of your life.

The first change is deciding what you are speaking for. A woman who speaks for approval will always be vulnerable to withdrawal. If the likes disappear, confidence collapses. If the comments turn hostile, she starts editing herself into safety. Speaking for approval turns expression into a performance contract that can be revoked at any time. Speaking for purpose is different. Purpose can survive silence from the crowd. Purpose can survive misunderstanding. Purpose can survive the fact that most people are not paying attention. The shift is choosing purpose over applause, because applause is not stable and it is not loyal.

The second change is deciding what you are willing to lose. This sounds harsh, but it is a clean way to gain control. If you are not willing to lose anything, you will say nothing true. If you are willing to lose attention, you can speak honestly. If you are willing to lose access to certain audiences, you can hold boundaries. If you are willing to lose the fantasy of being universally liked, you can stop managing the emotional needs of strangers.

Women are often taught that social safety depends on being acceptable. Online, that lesson becomes weaponized. The shift is accepting that some rejection is the price of being real, and that being real is what makes your work durable.

This shift also requires distinguishing audience from intimacy. Many women confuse them because platforms encourage it. People reply warmly, share personal stories, and create the feeling of closeness. Then the same people can become demanding, invasive, or punishing when you do not respond fast enough. Audience is not friendship. Audience is not family. Audience is a crowd that may contain good people, but it is still a crowd. Speaking on your own terms means deciding how close you allow strangers to get, and how much emotional labor you are willing to donate to people you will never meet.

Another part of the shift is learning the power of selective silence. Silence is not the absence of voice. Silence is the refusal to be pulled into someone else's agenda. Online conflict often functions like a dragnet. It is designed to catch your attention and force you into reaction, defense, or explanation. When you respond, you feed the system. When you explain yourself, you validate the frame. When you argue, you donate energy to people who are not building anything. Silence is not surrender. Silence is control. Silence is a decision to preserve your time for the work that actually matters.

This is where boundaries become a publishing strategy, not just a self-care concept. A woman who wants to stay visible long-term cannot treat every comment as a conversation invitation. She cannot treat every provocation as a challenge she must meet. She cannot treat every accusation as something that deserves a rebuttal. That posture turns visibility into a crisis cycle. Speaking on your own terms means choosing what you respond to and what you let die in the open air. Most noise wants oxygen. Starve it.

The shift also involves choosing the size of the room. Many women assume bigger is always better. More followers, more reach, more exposure. But bigger rooms produce more distortion. They collapse context faster. They invite more bad faith. They make you easier to target. Smaller rooms can be more stable, more sustainable, and more accurate.

A woman does not need the largest possible audience. She needs the right audience. She needs the audience that aligns with the work she is actually doing, not the audience that is there for entertainment.

There is also a shift in how you treat your voice. On most platforms, voice is treated like content. It is something you produce, package, and optimize. For women, that creates a constant temptation to shrink the message into whatever survives the feed. Over time, women start writing to avoid conflict rather than writing to tell the truth.

Speaking on your own terms means refusing to let the platform compress you. It means allowing yourself to be specific, even when specificity is inconvenient. It means refusing to translate everything into safe language for people who are not listening in good faith.

Decentralized tools can help here because they make it easier to build layered communication. You can have a public channel for publishing. You can have a quieter channel for trusted people. You can have a channel that exists mainly as continuity and redundancy. You are not forced into one all-purpose stage where every part of your life is visible at once. The shift is using platforms as tools instead of treating platforms as identity.

This is also where women should be honest about exposure. Visibility can be a weapon against isolation, but it can also be a weapon used against you. The more visible you are, the more searchable you are. The more searchable you are, the easier it becomes for strangers to frame you, clip you, or misrepresent you. This does not mean you should hide. It means you should choose visibility with the same seriousness you choose financial commitments. You are building an asset, and assets need protection.

For many women, the hardest part of speaking on their own terms is breaking the habit of over-explaining. Over-explaining is a survival technique. It is how women try to prove they are reasonable and worth listening to. But online, over-explaining becomes fuel. It gives bad-faith people more material to twist and more opportunities to keep you trapped. The shift is letting your statement stand without a fortress of disclaimers. Be clear, then be done.

The same principle applies to apology. Women apologize constantly online, even when they have done nothing wrong. They apologize for taking up space, being firm, or not responding quickly. Speaking on your own terms means keeping apologies for actual harm, not for existing. It means refusing to be trained into a smaller version of yourself. That is discipline.

This chapter explained how speaking on your own terms requires separating voice from availability, purpose from approval, and audience from intimacy. It showed that silence is not weakness, and that boundaries are structural control that keep women stable in environments designed to extract attention.

The shift is that you stop letting the platform define the relationship between you and your speech. You choose when to be visible and when to be silent. You choose who gets access to you, and you treat that choice as legitimate. Speaking on your own terms is a discipline that protects your voice so it can still exist tomorrow.

Conclusion

Nostr is not presented here as a new identity, a new tribe, or a better internet. It is presented as a tool that changes who holds continuity. For women, that distinction matters because modern communication systems are not neutral. They shape what gets seen, what gets punished, what gets rewarded, and what quietly disappears.

Across these chapters, the aim has been practical. To name the tradeoffs clearly. To show what decentralization can and cannot do. To describe what changes when identity becomes portable and publishing is less dependent on a single platform's permission. That kind of understanding does not require belief. It requires discernment.

A woman does not need a perfect network. She needs options. She needs channels that do not collapse when policies shift, algorithms tighten, or visibility becomes conditional. She needs the ability to step back without losing herself, and to publish without being trapped in constant performance.

Nostr does not make the world kinder. It does not prevent harassment, misunderstanding, or conflict. What it offers is a different structure, one where control is not concentrated in a single corporate gate. That structure demands boundaries, selectivity, and restraint. But it also restores something women are often pressured to surrender. The ability to decide the terms of their own presence.

EATMS Productions

EATMS Productions publishes books and videos for women navigating modern systems.

One woman spent years reading, watching, photographing, writing, gardening, thinking, getting pissed off, walking away from conventional cultural institutions, watching technology dismantle some of the gates that had kept people outside, and eventually realized:

Fine. I'll build the fucking thing myself.

This book is part of what she built.

Across different subjects, series, authors, characters, and formats, EATMS is a single connected body of work examining power, technology, economics, culture, history, and the systems shaping women's lives. Different books speak in different voices. Propaganda Paige wanders through the archive. New work keeps getting added.

80+ books. All PDFs free.

Free digital editions:
eatms.gumroad.com
Full catalog:
eatms.me/catalog
EATMS Productions:
eatms.me

Please feel free to burn part or all of this book, safely, as an effigy.

New Books by EATMS

Non-Fiction

Media Literacy for Women: Truth Will Rise
Journey Live in Houston, 1981: A Middle-Aged Midwestern Woman's Story
Heaven or Las Vegas
The Last Mall
Friday Night at the Video Store: The Movies We Found
She Wasn't Waiting to Be Rescued
The World Was Supposed to Be Better Than This
The Singularity
How Money Killed All the Good Artists
AI Ate the Electric Bill: Data Centers, Privacy, & Who Pays
Men Should Pause: A Self-Help Guide to Male Menopause
MAHA, Ha Ha: You Are What They Let You Afford
Stop Smiling at Nazis
He's So Cute: What Makes a Man Worth Liking
NSPM-7 & Gen X: How to Survive in Post America
What to Remember: How to Survive in Post America

Fiction

I Sit Low on the Floor: A Novel About Staying Low
Stoogey Old Hag & the Cat
The Girls at Bellwether Mall
The Girls Below Bellwether
When Bellwether Opens Again

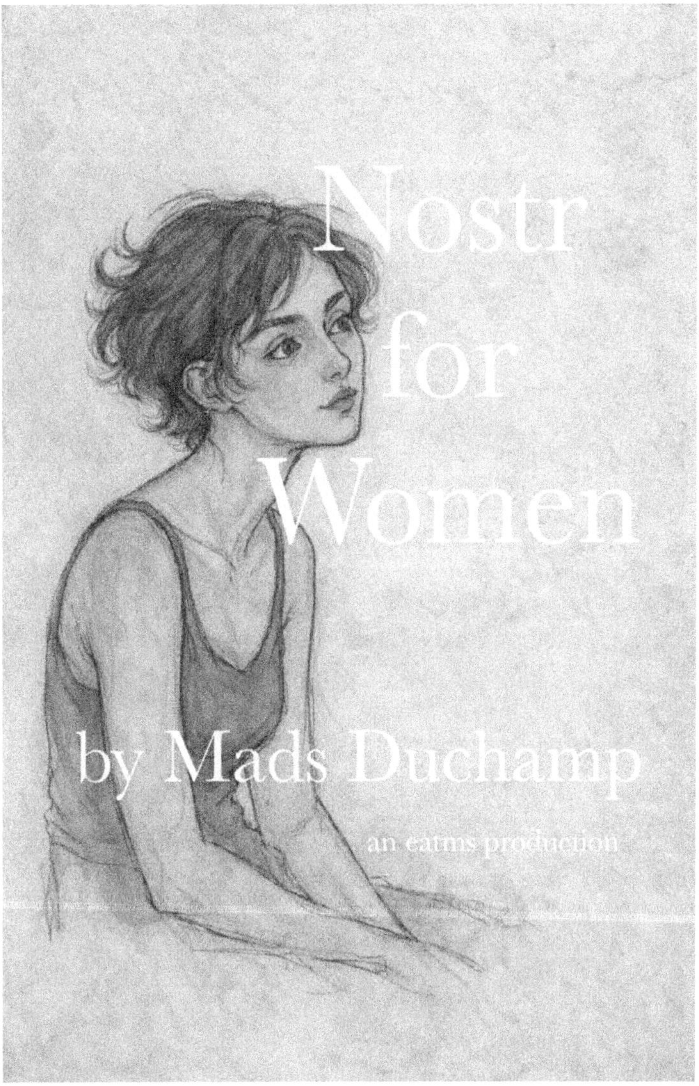
Nostr
for
Women
by Mads Duchamp
an eatms production

Made in United States
Orlando, FL
05 November 2026

90230135R00069